CLIMATE CHANGE IN CONTRASTING RIVER BASINS
Adaptation Strategies for Water, Food and Environment

CLIMATE CHANGE IN CONTRASTING RIVER BASINS
Adaptation Strategies for Water, Food and Environment

Edited by

Jeroen C.J.H. Aerts

Institute for Environmental Studies
Free University
Amsterdam
The Netherlands

and

Peter Droogers

FutureWater
Arnhem
The Netherlands
and
International Water Management Institute
Colombo
Sri Lanka

CABI Publishing

CABI Publishing is a division of CAB International

CABI Publishing
CAB International
Wallingford
Oxfordshire OX10 8DE
UK

CABI Publishing
875 Massachusetts Avenue
7th Floor
Cambridge, MA 02139
USA

Tel: +44 (0)1491 832111
Fax: +44 (0)1491 833508
E-mail: cabi@cabi.org
Website: www.cabi-publishing.org

Tel: +1 617 395 4056
Fax: +1 617 354 6875
E-mail: cabi-nao@cabi.org

A catalogue record for this book is available from the British Library, London, UK.

Library of Congress Cataloging-in-Publication Data

Climate change in contrasting river basins : adaptation strategies for water, food, and environment /edited by J. Aerts, P. Droogers.
    p. cm.
  Includes bibliographical references and index.
  ISBN 0-85199-835-6 (alk. paper)
  1. Climatic changes. 2. Crops and water. 3. Crops and climate. I. Aerts, J. (Jeroen)
II. Droogers, Peter, 1961- III. Title.

  S600.7 C54C66 2004
  630'2'516--dc22   2004000968

ISBN 0 85199 835 6

Typeset by Servis Filmsetting Ltd, Manchester
Printed and bound in the UK by Biddles Ltd, King's Lynn

Contents

Contributors

Aerts, J., *Institute for Environmental Studies (IVM), Faculty of Earth and Life Sciences, Vrije Universiteit Amsterdam, De Boelelaan 1087, 1081 HV Amsterdam, The Netherlands.*

Agha Alikhani, M., *College of Agriculture, Department of Irrigation and Drainage, Tarbiat Modarres University, PO Box 14115-336, Tehran, Iran.*

Andah, W., *Water Research Institute, PO Box 38, Achimota, Ghana.*

Biney, C.A., *Water Research Institute, PO Box 38, Achimota, Ghana.*

Bouwer, L.M., *Institute for Environmental Studies (IVM), Faculty of Earth and Life Sciences, Vrije Universiteit Amsterdam, De Boelelaan 1087, 1081 HV Amsterdam, The Netherlands.*

Chevnina, E.V., *Faculty of Geography, Lomonosov Moscow State University, Leninskiye Gory, Moscow 119992, Russia.*

de Ruyter van Steveninck, E., *UNESCO-IHE Institute for Water Education, PO Box 3015, 2601 DA Delft, The Netherlands.*

Douben, K.J., *UNESCO-IHE Institute for Water Education, PO Box 3015, 2601 DA Delft, The Netherlands.*

Droogers, P., *FutureWater, Eksterstraat 7, 6823 DH Arnhem, The Netherlands.*

Gieske, A., *International Institute for Geo-Information Science and Earth Observation (ITC), PO Box 6, 7500 AA, Enschede, The Netherlands.*

Guttman, H., *Mekong River Commission Secretariat, PO Box 1112, 364 Preah Monivong Boulevard, Phnom Penh, Cambodia.*

Hoanh, C.T., *International Water Management Institute (IWMI), PO Box 2075, Colombo, Sri Lanka.*

Hoogeveen, J., *FAO, Land and Water Development Division, Viale delle Terme di Caracalla, 00100 Rome, Italy.*

Huber-Lee, A., *Stockholm Environment Institute, 11 Arlington Street, Boston, MA 02116-3411, USA.*

Jayatillake, H.M., *Irrigation Department, PO Box 1138, Colombo – 07, Colombo, Sri Lanka.*

Klein, H., *UNESCO-IHE Institute for Water Education, PO Box 3015, 2601 DA Delft, The Netherlands.*

Lasage, R., *Institute for Environmental Studies (IVM), Faculty of Earth and Life Sciences, Vrije Universiteit Amsterdam, De Boelelaan 1087, 1081 HV Amsterdam, The Netherlands.*

Loeve, R., *FutureWater, Eksterstraat 7, 6823 DH Arnhem, The Netherlands.*

Mannaerts, C., *International Institute for Geo-Information Science and Earth Observation (ITC) PO Box 6, 7500 AA Enschede, The Netherlands.*

Massah, A.R., *College of Agriculture, Department of Irrigation and Drainage, Tarbiat Modarres University, PO Box 14115-336, Tehran, Iran.*

McCluskey, A., *University of Colorado, Department of Civil and Environmental Engineering, PO Box 428, Boulder, CO 80306, USA.*

Mohammadi, K., *College of Agriculture, Department of Irrigation and Drainage, Tarbiat Modarres University, PO Box 14115-336, Tehran, Iran.*

Morid, S., *College of Agriculture, Department of Irrigation and Drainage, Tarbiat Modarres University, PO Box 14115-336, Tehran, Iran.*

Purkey, D., *Natural Heritage Institute, 2140 Shattuck Avenue, 5th Floor, Berkeley, CA 94704, USA.*

Runkle, B., *Stockholm Environment Institute, 11 Arlington Street, Boston, MA 02116-3411, USA.*

Savoskul, O.S., *Faculty of Geography, Lomonosov Moscow State University, Leninskiye Gory, Moscow 119992, Russia.*

Strzepek, K., *University of Colorado, Department of Civil and Environmental Engineering, PO Box 428, Boulder, CO 80306, USA.*

Werners, S., *Climate Change and Biosphere Research Centre (CCB), Wageningen UR, PO Box 47, 6700 AA Wageningen, The Netherlands.*

Yates, D., *Research Applications Program, National Center for Atmospheric Research, 3450 Mitchell Lane, Boulder, CO 80301, USA.*

Young, C., *Research Applications Program, National Center for Atmospheric Research, 3450 Mitchell Lane, Boulder, CO 80301, USA.*

Yu, W., *Stockholm Environment Institute, 11 Arlington Street, Boston, MA 02116-3411, USA.*

van Dam, J., *Wageningen University, Nieuwe Kanaal 11, 6709 PA Wageningen, The Netherlands.*

van Deursen, W., *Carthago Consultancy, Oostzeedijk Beneden 23a, 3062 VK Rotterdam, The Netherlands.*

van de Coterlet, G.M., *Institute for Environmental Studies (IVM), Faculty of Earth and Life Sciences, Vrije Universiteit Amsterdam, De Boelelaan 1087, 1081 HV Amsterdam, The Netherlands.*

van de Giesen, N., *Center for Development Research (ZEF), Bonn University, Walter-Flex Straße 3, D-53113 Bonn, Germany.*

Foreword

The issue of adaptation to climate change has become increasingly important in recent years as it is becoming evident that the impacts of human-induced climate change are no longer just a long-term possibility, but are actually a medium-reality probability.

One of the sectors that is likely to be impacted quite severely is the large freshwater systems and river basins. It is therefore important for water-sector policy makers, planners and managers to take into account the possible impacts of climate change on the systems they are managing. In order to do so effectively they will need appropriate tools and methods to enable them to model the potential impacts of climate change and test appropriate adaptation options. This book describes one such effort to develop and test a tool called the Adaptation Methodology for River Basins (AMR). The framework is largely built on existing knowledge, but adapts it to the needs of the water managers. Adaptation in water management is a new issue and adaptation at the basin level is especially difficult. However, it is important that adaptations be considered at the basin level before single-country based projects are developed, in order to avoid making the whole system even more vulnerable to climate change impacts. The book also focuses on adaptations for alleviating impacts for food security and environmental values, with special attention to hydrological and food modelling. I am sure that the book will prove to be a valuable tool for water-sector managers and planners dealing with large river basins to enable them to take into account the potential impacts of climate change and develop adaptation options.

Saleemul Huq
Director, Climate Change Programme
International Institute for Environment and Development (IIED)

1 Adaptation for Regional Water Management

JEROEN AERTS[1] AND PETER DROOGERS[2]

[1]*Institute for Environmental Studies, Vrije Universiteit Amsterdam, Amsterdam, The Netherlands; [2]FutureWater, Arnhem, The Netherlands*

Introduction

Dealing with climate change and climate variability is generally considered to be one of the largest challenges for the coming decades, on all geographical scales, across all economic sectors. Water managers see themselves confronted with a continuous stream of increasingly credible scientific information on the potential magnitude of climate change and climate variability and the vulnerability of water resources to its impacts. The urgency to take action is more apparent than ever, yet clear guidance on exactly how to respond to the challenge of climate change is lacking, especially at the river basin level. A consensus, however, is emerging around the viewpoint that it is not only necessary to mitigate climate change by reducing greenhouse gas emissions and enhancing carbon sinks, but also to adapt to the inevitability of climate change by preparing for impacts and reducing vulnerability (Pielke, 1998; Kane and Shogren, 2000).

The Intergovernmental Panel on Climate Change (IPCC) states that changes in precipitation are likely to have a major impact on the hydrological cycle and, subsequently, on the environment and on food production (Arnell *et al.*, 2001). Currently 800 million people suffer from hunger, among them 200 million children under 5 years of age. It is estimated that by 2025 cereal production will have to increase by 38% to meet world food demands. Climate change (CC) is expected to amplify climate variability (CV) and hence the occurrence of extreme events such as floods and droughts. For example, it is expected that cereal production in sub-Saharan Africa will decrease, with 2–3% of this decrease being due to increased climate variability (FAO, 2003). Allocating more water to agriculture, however, is not always the answer to these problems, since water is used for other sectors, such as nature and hydropower.

The ADAPT project

This book is the result of the ADAPT project, which focused on developing regional adaptation strategies for water, food and the environment in river basins across the

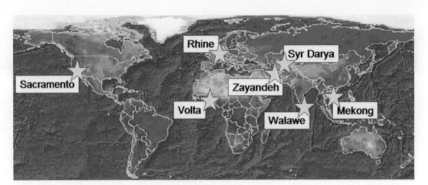

Fig. 1.1. The seven basins included in the ADAPT project.

world. The projected impacts of CC on water resources underline the necessity of water managers to seek new and sustainable water allocation measures that address the potential impact of CC. Response or adaptation in water management to ensure food security under changing climatic conditions is a difficult process, where complex trade-offs have to be made across different policy objectives and hence requires input from all stakeholders involved in this process. Despite its complexity, adaptation increasingly receives attention in policy making as a complementary coping mechanism to mitigation. Adaptation is explicitly addressed in several policy documents, such as in article 10 of the Kyoto Protocol (UNFCCC, 1997), where 'parties are further committed to promote and facilitate adaptation and deploy adaptation technologies to address climate change'. Furthermore, the process of adaptation is not new. Throughout history people have adapted to changing or extreme climate conditions. Especially in the water sector, people can learn from adaptation experiences in the past (Tol *et al.*, 1998).

As impacts primarily take place at the regional (river basin) or local scales, adaptations similarly should be sought at those scales (IPCC, 2001b). The basins for the ADAPT project were selected within dry and wet areas and located in developed and developing countries. The rationale for formulating this project was that exchanging knowledge on adaptation drawn from very different regions would accelerate scientific innovation in this relatively unexplored field. The seven basins selected are (see Fig. 1.1):

- Mekong, South-east Asia;
- Rhine, Western Europe;
- Sacramento, USA;
- Syr Darya, Central Asia;
- Volta, Ghana;
- Walawe, Sri Lanka; and
- Zayandeh, Iran.

The main goal of the ADAPT project was to develop a generic methodology for river basins (called Adaptation Methodology for River Basins, AMR) that allows the development and assessment of adaptation strategies for alleviating food and environmental impacts (Fig. 1.2). The methodology puts stakeholders in a central role in the adaptation process and iteratively addresses the following steps.

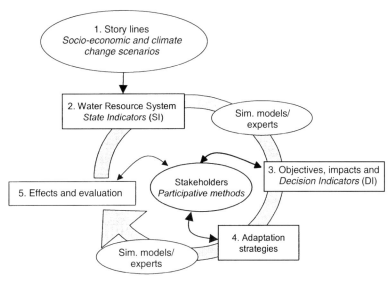

Fig. 1.2. An overview of the adaptation framework for river basins (AMR) followed by all case study areas in the ADAPT project (Aerts *et al.*, 2003).

- Derive storylines for each basin, such as socio-economic developments, and climate change and variability projections per basin.
- Select a set of models at basin and field scale capable of simulating hydrology and food production.
- Assess CC/CV impacts on regional water management by comparing future simulations with baseline references with respect to the environment and food security.
- Define adaptation strategies for water managers to respond to climate change.
- Evaluate adaptation strategies.

Purpose of this book

This book can be seen as a handbook for regional water management to develop and evaluate adaptation strategies to climate change and climate variability. For this, the generic AMR methodology is first described in this chapter, and next explained for seven case studies. These basin case studies (Chapters 5–11) followed this common methodology and each of the framework steps is addressed in the different basin chapters of this book. Apart from the basin chapters, several supporting chapters provide in-depth supporting information. Chapter 2 discusses the use of climate change scenarios as provided by the IPCC and, more specifically, how these scenarios can be used for regional studies. Chapters 3 and 4 describe in more detail the possible consequences of CC and CV for food security and environmental quality, respectively. Chapter 12 integrates the findings of the basin studies and compares these findings with global trends in climate change related to food security. Finally, Chapter 13 provides a summary of the experiences encountered during this project and provides the reader with some key findings that should be addressed in new regional adaptation studies.

Climate Change and Water Resources

Strong scientific evidence indicates that the average temperature of the earth's surface is increasing due to greenhouse gas emissions. The average global temperature has increased by about 0.6° since the late 19th century (IPCC, 2001a). The latest IPCC scenarios project temperature rises of 1.4–5.8°C, and sea level rises of 9–99 cm by 2100 (Figs 1.3 and 1.4; IPCC, 2001a). Warming and precipitation are expected to vary considerably from region to region as the greatest increase in temperature is expected in the Northern Hemisphere. Changes in climate averages, coupled with changes in the frequency and intensity of extreme weather events, are likely to have major impacts on natural and human systems. Effects on the world's poor in tropical and sub-tropical areas are likely to be disproportionately large, in particular since their potential to adapt to such changes is low (Smit *et al.*, 2001).

Changes in the cycling of water between land, ocean and atmosphere can have significant impacts across many sectors of the economy, society and environment. Consequently, there are many studies that focus on the effects of CC on the hydrological cycle and the related availability of water resources for human and environmental use. The majority of these studies have focused on changes in the water balance. Other, but fewer, studies focused on the impacts of CC on water resources in terms of the reliability of the water supply, the risk of flooding or on exploring possible adaptation strategies (Arnell *et al.*, 2001; Kabat and van Schaik, 2003).

Hydrological impacts

Climate change is only one of the pressures facing the hydrological system and water resources. Other global changes, such as population growth, pollution, land use changes and land management, also have a profound impact on the hydrological cycle. In general, there is an increasing move towards sustainable water management and increasing concern for the impacts of global change on the water resources system. Recent initiatives to address these issues include, for example, the 'Dublin Statement' in 1992, which urges the sustainable use of water; and activities by the World Water Council, which led to a vision for a 'water secure world' (Cosgrove and Rijsberman, 2000) and a report that addresses the need for water managers to better prepare for change in climate (Kabat and van Schaik, 2003).

The IPCC has assessed the major recent studies into the effects of CC on hydrology in its Third Assessment Report (TAR) (Arnell *et al.*, 2001). IPCC found that most hydrological studies on the effects of CC have concentrated on streamflow and runoff (streamflow is water within a river channel, whereas runoff is the amount of precipitation that does not evaporate). Changing patterns in runoff are consistent with those identified for precipitation. However, in large parts of eastern and northwestern Europe, Canada and California, a major shift in streamflow from spring to winter has been associated with a change in precipitation. It appears that in these areas

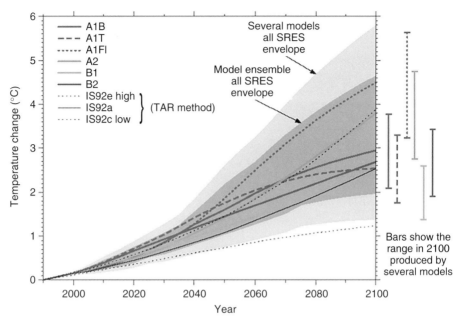

Fig. 1.3. Global temperature projections, according to different scenarios. (From IPCC, 2001a.)

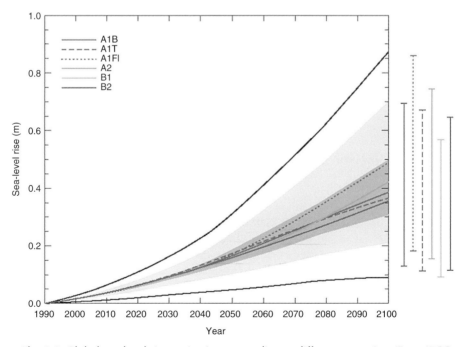

Fig. 1.4. Global sea-level rise projections according to different scenarios. (From IPCC, 2001a.)

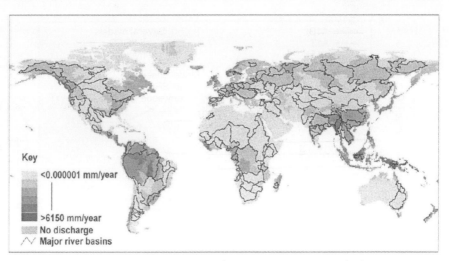

Fig. 1.5. A map of long-term average annual water resources (MAR) by basin, calculated by the WaterGAP2 model. (From Smakhtin *et al.*, 2004.)

precipitation will fall as rain, rather than snow, in winter periods (e.g. Lettenmaier *et al.*, 1999; Middelkoop *et al.*, 2001). In colder regions, no significant changes have been observed. Figure 1.5 shows a global map of long-term average annual water resources per major river basin (Smakhtin *et al.*, 2004).

Increased temperatures generally result in an increase in potential evapotranspiration. In dry regions, potential evapotranspiration is driven by energy and is not constrained by atmospheric moisture contents. In humid regions though, atmospheric moisture content is a major limitation to evapotranspiration. Studies show increases in evapotranspiration with increased temperatures. However, models using equations that do not consider all meteorological controls may be very misleading. Vegetation plays an important role in evaporation by intercepting precipitation and by determining the rate of transpiration. Higher CO_2 concentrations may lead to increased water use efficiency (WUE, or water use per unit biomass), implying a reduction in transpiration. Higher CO_2 concentrations may also be associated with increased plant growth, compensating for increased WUE – thus, plants may acclimatize to higher CO_2 concentrations. The actual rate of evaporation is constrained by water availability.

Most increases in greenhouse gasses (GHGs) are associated with reduced soil moisture in the Northern Hemisphere summers (IPCC, 2001a). This is the result of higher winter and spring evaporation, caused by higher temperatures and reduced snow cover, and lower rainfall during the summer. It appeared that a lower water-holding capacity of the soil results in greater sensitivity to CC. Furthermore, increased winter rainfall in the Northern Hemisphere may result in increased groundwater recharge. However, increased temperatures may increase evaporation, which leads to longer periods of soil water deficits.

In general, however, it is difficult to identify trends in hydrological data (Fig. 1.6). First of all, records are short, and monitoring stations are continuing to be closed in many countries. An alternative is the use of remote sensing to assess runoff. Secondly,

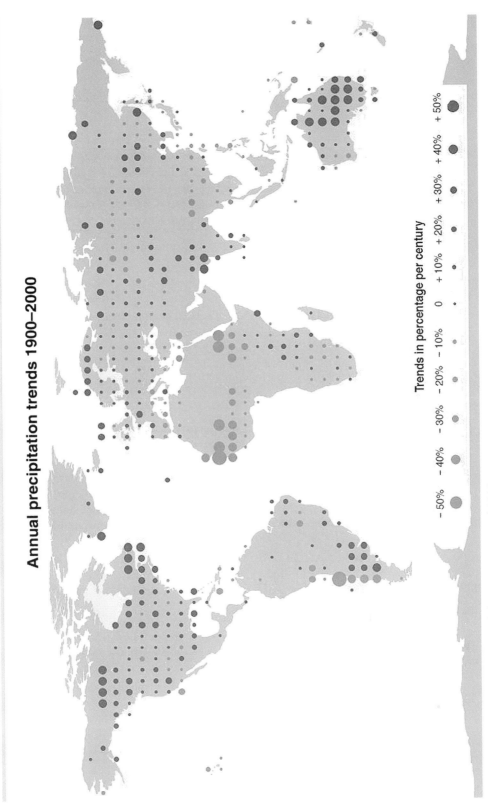

Fig. 1.6. Annual precipitation trends between 1900 and 2000.

trends in the records of stream flow are obscured by interannual and decadal scale climate variability and non-climatic factors, such as land use change and various river management measures. Some general implications described by the IPCC (Arnell *et al.*, 2001), however, are:

- in systems with large reservoirs, changes in resource capacity may be proportionally smaller than changes in riverflows;
- potential effects of CC must be considered in the context of changes in water management – CC changes may have little effect on the water resources as compared to changes in water management over a period of 20 years; and
- the implications of CC are likely to be the greatest in systems that currently are highly stressed.

Impacts on extreme events

Relatively few studies have examined CC effects on flooding frequencies. The main reason is that the generally available general circulation model (GCM) scenarios are monthly averages, which are not suitable in the study of small-scale, short duration events. A flood frequency study was conducted by Mirza (2002) in South Asia. Here, according to four GCM scenarios, the flood discharges in the Ganges Brahmaputra Meghna (GBM) basin could increase by 6–19%.

Droughts are even more difficult to define in quantitative terms compared to floods, since changes in water resources management have a relatively large effect on drought as well as climatic and hydrological inputs. The effects of drought are often expressed with water resources stress indicators. These include, for example, 'amount of water available per person' and 'ratio of volume withdrawn to volume available'. Projections show that 0.5 billion people could see increased water resources stress by 2020 as a result of CC. Case studies show that the impacts of different demands and operational assumptions by 2050 are greater than, or of similar magnitude to, the potential impacts of CC (e.g. Lettenmaier *et al.*, 1999).

Regional differentiation in impacts

The impacts of CC on hydrology are usually estimated by defining scenarios for changes in climatic inputs through a hydrological model. These scenarios can be of two basic types: (i) synthetic scenarios, consisting of assumed changes in temperature and precipitation, using generally available datasets or a weather generator; or (ii) outputs from previously mentioned general circulation models (GCMs) are also increasingly used.

The use of GCM output data, however, suffers from three major problems. First, GCMs are often not able to simulate the current observed regional and local climate. Secondly, the spatial scale is too coarse to be directly used in hydrological models that simulate changes at a much lower spatial and temporal scale. Thirdly, different models project different changes in precipitation. The latter is mainly because changes in the hydrological cycle due to CC are more difficult to simulate than potential impacts through temperature changes because precipitation observations are less complete and the physical constraints are weaker (e.g. Dvořák, 1997; Allen and Ingram, 2002).

Regional precipitation patterns have been modelled by GCMs under different climate change scenarios (see also Chapter 2). The results and the analysis of cross-model consistency for two different emission scenarios (called A2 and B2) in regional precipitation change have been assessed by IPCC (2001a).

Climate Change and Water Use

Water demand is a synonym for human and environmental 'water requirements'. There are in-stream demands (no withdrawals, e.g. hydropower generation, navigation) and off-stream demands (withdrawals). Off-stream demands can be either consumptive (e.g. irrigation) or non-consumptive (water is returned to the river).

Agricultural use is the largest consumer of water around the world, accounting for 67% of all withdrawals and 79% of all water consumed (FAO, 2003). Municipal or domestic uses account for 9% of withdrawals. It is expected that water withdrawals would exceed 40% of annual water availability comparing 2025 and 1995 values (Alcamo and Heinrichs, 2001). The greatest rates are projected in developing countries, e.g. in Africa and the Middle East (without taking CC into account). Water withdrawals are expected to fall in developed countries because of water pricing, for example. Industrial water withdrawals account for 20% of all withdrawals. Without CC, these withdrawals will increase, and are concentrated largely in Asia, Latin America and Africa (IPCC, 2001b).

The expected effects on agricultural use are: (i) a change in field-level climate may alter the need for and timing of irrigation. Increased dryness may lead to increased demand, but demand could be reduced if soil moisture content rises at critical times. (ii) Higher CO_2 concentrations would lower plant stomatal conductance, hence increase the WUE, but this may be offset to a large extent by increased plant growth.

An important impact of CC and CV is 'drought', when rainfall drops below the long-term average, and the largest regional reduction in cereal production is therefore expected in Africa (2–3%). At higher latitudes, increased temperatures can lengthen the growing season and ameliorate cold temperature effects on growth. In warmer mid-latitude environments, adverse effects could include increased pests and crop diseases, soil erosion, increased flooding, desertification and reduced water resources for irrigation (Fischer *et al.*, 2001).

When studying the effects of CC on water for food security, related issues have to be considered. First, the effects due to CC are small compared to economic and technological growth. Secondly, it is expected that a rise in atmospheric CO_2 can also be a positive factor in tree and crop growth and biomass production. It stimulates photosynthesis (the so-called CO_2 fertilizer effect) and improves water use efficiency (Bazzaz and Sombroek, 1996).

Adaptation

Adaptation increasingly receives attention in policy making as a complementary coping mechanism to mitigation. Adaptation is explicitly addressed in several policy documents, such as the declarations of the UNFCCC (1992), where Article 4.1b states that 'parties are committed to formulate and implement national, and where appropriate, regional programmes containing measures to facilitate adequate adaptation to Climate change'. This has been strengthened in article 10 of the Kyoto Protocol (UNFCCC, 1997), where 'parties are further committed to promote and facilitate adaptation and deploy adaptation technologies to address Climate change'.

As stated in the Introduction, the process of adaptation is not new. Throughout history, people have adapted to changing or extreme climate conditions. Think of the development of both Dutch and Bengal villages on dykes and man-made hills in flood-prone areas, or the development of irrigation and reservoir networks in the desert areas of California and Iran. In these cases, people have adapted their way of living to extreme environmental conditions. Within the context of anthropogenic climate change, however, the question is how robust these adaptations are to future, unknown extreme conditions.

Different definitions of adaptation exist. Burton *et al.* (1998, p. 66) describe adaptation as 'responses to climate change that may be used to reduce vulnerability', where vulnerability is defined as susceptibility to harm or damage potential. Furthermore, adaptation considers such factors as the ability of a system to cope or absorb stress or impacts and to 'bounce back' or recover. Adaptation can also refer to actions designed to take advantage of new opportunities that may arise as a result of climate change. Pittock and Jones (2000) define adaptation as a response to climate change that seeks to maintain viability by maximizing benefits and minimizing losses.

In order to study the development and evaluation of adaptation strategies for river basins, ADAPT will assess the necessary ingredients of an adaptation strategy and aim to thoroughly understand the system to which the adaptation strategy will be applied. Smit *et al.* (1999) specify a number of elements that need to be addressed in any scientific adaptation research. These elements, posed as questions, are: (i) who or what adapts? (ii) Adapts to what? (iii) How does adaptation occur? A possible fourth question relates to the quality of adaptation: How good is the adaptation? (Fig. 1.7).

Who or what adapts?

The first element refers to the system definition: who or what adapts? Is it an individual or a community, a region or a nation? Are we considering a species or an ecosystem? A system definition is important for studies where impacts are assessed, with and without adaptation and to determine its vulnerability.

Once the system has been defined, the next step is to characterize the system. These characteristics should allow for assessing the prospects of planned adaptation and adaptive capacity (the latter is also referred to as autonomous adaptation) of the system under survey (Smit *et al.*, 1999). Important system characteristics are: sensitivity, vulnerability, susceptibility, coping range, critical levels, adaptive capacity, stability, robustness, resilience and flexibility (e.g. Klein and Tol, 1997; Smit *et al.*, 1999; Reilly and Schimmelpfennig, 2000).

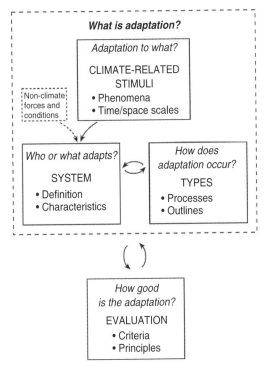

Fig. 1.7. Main elements for studying adaptation strategies. (From Smit *et al.*, 1999.)

The most broadly used terms for a system's characterization are: *sensitivity, adaptive capacity* (or adaptability) and *vulnerability* (Smit *et al.*, 1999). Sensitivity can be defined as the degree to which a system is affected by or responsive to climate stimuli. Adaptive capacity is the extent to which sectors, regions and communities ('the system') are able to adapt to climate change impacts. Adaptive capacity reflects the notion that the existence of adaptation options does not necessarily mean that each vulnerable sector, region or community has access to these options or is in a position to implement them. In other words, it is the capacity to adapt rather than the availability of adaptation options that determines the degree of resilience to climate change (Smit *et al.*, 2001). Options to increase adaptive capacity include: increasing wealth, scientific understanding, technology and flexibility (e.g. through developing early warning systems). Finally, the vulnerability of a given system or society is a function of its physical exposure to climate change effects and its ability to adapt to these conditions (Kelly and Adger, 2000; Smit *et al.*, 2001).

The IPCC (Smit *et al.*, 2001) further states that by assessing differences in vulnerability among regions and groups and by working to improve the adaptive capacity of those regions and groups, planned adaptation can contribute to equity considerations of sustainable development and may contribute to alleviating poverty in developing countries. Hence, vulnerability is a function of sensitivity and adaptive capacity and its concept is increasingly seen as key input for developing and evaluating adaptation strategies (IPCC, 1998).

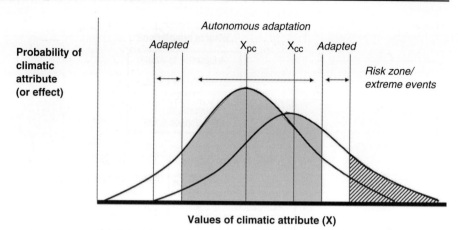

Fig. 1.8. A shift of climate variable X_{pc} in terms of its mean and frequency distribution results in a future distribution X_{cc}. The figure shows how the current adaptation zones are insufficient in the future, since the risk zone (shaded zone) increase dramatically.

Adaptation to what?

The second element in adaptation studies refers to determining the climate stimuli relevant to the system and to relate climate stimuli to the sensitivity of the system. Climate conditions, such as temperature and precipitation, can be classified into three temporal categories: (i) long-term changes; (ii) inter-annual or decadal changes; and (iii) isolated extreme events. This notion is important, since current research emphasizes the importance of studying both gradual climate change and climate variability, including extreme events. Spatial characteristics of climate stimuli also play a role, as projected regional climate change can be very different from what is simulated for the globe (Smit, 1993; Tol, 1996).

The most important climate stimuli that influence the hydrological cycle of a river basin and hence the availability of water resources are temperature and precipitation. It is, therefore, important to derive some quantitative projections of expected temperature and precipitation changes in the future. From these projections, impacts or effects can be determined. It is also important to make a distinction between changes in the frequency of extreme events versus gradual climate changes. This can be illustrated with Fig. 1.8. It shows the frequency distribution of variable climate X_{pc} (e.g. precipitation) with adaptation ranges. That is, currently adaptation measures have been implemented to cope with extreme events within those ranges. However, a change in the mean and frequency of a climate variable will cause a shift in the risk zone, or the zone that is not covered through either autonomous or planned adaptation. The risk zone in the future is considerably larger than in the current situation (Smit *et al.*, 1999; Kabat and van Schaik, 2003).

How does adaptation occur?

This element refers to the process of developing and implementing adaptation and the forms of adaptation. Figure 1.9 summarizes the general types of adaptation and some examples, which are differentiated according to *timing* (anticipatory versus reactive) and

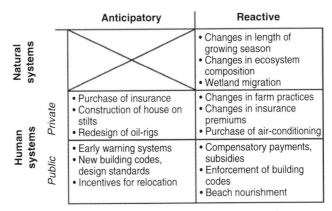

		Anticipatory	Reactive
Natural systems			• Changes in length of growing season • Changes in ecosystem composition • Wetland migration
Human systems	*Private*	• Purchase of insurance • Construction of house on stilts • Redesign of oil-rigs	• Changes in farm practices • Changes in insurance premiums • Purchase of air-conditioning
	Public	• Early warning systems • New building codes, design standards • Incentives for relocation	• Compensatory payments, subsidies • Enforcement of building codes • Beach nourishment

Fig. 1.9. Types of adaptation to climate change, including examples. (From Smit *et al.*, 2001.)

human versus natural. Anticipatory adaptation is also referred to as preventive adaptation (thus implemented 'before the event'). Human-induced adaptation is also referred to as planned adaptation and can be differentiated according to the type of agents – private enterprises such as producers and industries versus public governmental organizations. Other characteristics of adaptation options include intent, duration, form and type (Klein and MacIver, 1999; MacIver and Dallmeier, 2000; Smit and Skinner, 2002).

Most adaptations are modifications to existing practices and public policy decision-making processes such as those already developed in the agriculture and water sector. Therefore, there is a need for a better understanding of the relation between potential adaptations and existing practices (Reilly and Schimmelpfennig, 2000). Kabat and van Schaik (2003) describe an extensive compendium of possible adaptation measures. These measures have formed the basis for developing adaptation strategies studied within this book.

How good is the adaptation?

Adaptation research warns that most impact studies do not consider adaptation as an integral part of their assessment. Without assessment of adaptations, these impact studies could well overestimate the potential negative effect of climate change (Burton *et al.*, 1998). It is, therefore, suggested to include adaptation options iteratively within impact assessment studies by first identifying potential impacts without adaptation and next simulating impacts including adaptation. The potential set of adaptation options that results from this iterative process can finally be evaluated using a number of evaluation methods, e.g. sensitivity analysis, scenario analysis and multi-criteria analysis. Another evaluation technique is a cost–benefit analysis, but estimations of adaptation costs versus the losses without adaptation is still a relatively unexplored scientific field (Leary, 1999; Tol *et al.*, 1998).

Challenges for adaptation research

Efforts are required to better understand the human response to climate change, as it is likely that socio-economic systems respond the most to extreme realizations of climate change (Yohe and Dowlatabadi, 1999). Therefore, adaptation research needs to consider socio-economic scenarios, although it may affect the uncertainty of the result. Also, it is important to distinguish between a prediction or estimation of the *effect* of adaptations and, on the other hand, a normative study that focuses on the *evaluation* of adaptations. The latter requires information from the impact study to derive a set of feasible adaptations and to evaluate this set of adaptations. Policy response studies that incorporate evaluation of adaptation options are relatively new.

With a greater focus on adaptation, the debate on the impacts of CC will not stand in the way of effective mitigation, as most adaptations make sense under any CC scenario (Pielke, 1998). Policy options proposed as adaptation measures to reduce negative impacts of climate change that would be justified even in the absence of climate change are referred to as 'no regret' measures (IPCC, 2001b).

As described above, costs of adaptation are rarely studied and even less is known about the benefits. Most studies focus on total damage costs, including adaptation, and not on avoided damages through adaptation (e.g. Zeidler, 1997). It is estimated that, globally, adaptation costs only comprise about 7–10% of the total damage costs (Tol *et al.*, 1998). Another issue for further study is to estimate transition costs, since most adaptation and impact studies assume equilibrium now and in the future. However, climate continues to change, as do the impacted systems. The optimum level of adaptation minimizes the combined costs of adaptation and residual negative effects, with the most cost-effective steps taken first. Factors that affect adaptive capacity itself include: institutional capacity, wealth, planning time, scale, etc. (Tol *et al.*, 1998).

Finally, more research on the *timing* of adaptation is required. In this respect, Burton *et al.* (1998) point out that adaptation in socio-economic sectors is easier when investments relate to activities with a shorter product cycle. For example, different cropping methods can be adjusted every year. But a forest has a life-cycle of decades. Dams are even costlier to reconstruct in order to meet new climate conditions.

Important to realize is that adaptation measures are not only related to climate change. They can be a response to other internal (= manageable) and external (= less manageable) stressors. Examples of these stressors that should be taken into account are: land use change, population growth, increased competition between sectors (urban, industry, agriculture, nature), power generation, transboundary water allocation, environmental concerns, etc. The term 'no regret strategy' is therefore often used, since implementing an adaptation may solve a problem due to climate change in the future. But if not, it may solve other even bigger problems.

Adaptation and agriculture

It is projected that the paramount issue in changes in precipitation will be the increase in extremes rather than a long-term change in average precipitation. Increasing the buffer capacity is therefore the appropriate adaptation measure, where buffer capacity should be considered in terms of increased water storage (reservoirs, soil water, ground water) but also increased economic (savings/loans) and food buffer capacity. Essential is that an increase in extremes includes an increase in successive years of dry or wet periods, which are very difficult to overcome for poor people. A poor farmer might overcome a 1-year drought followed by a normal year, but a period of 2 or more years of drought, even followed by a longer period of normality, will be catastrophic to this farmer.

During the last decade, research has focused on the impacts of climate change on agricultural production (Parry *et al.*, 1999; Fischer *et al.*, 2001; FAO, 2003). These are studies at the global scale and only recently have impact studies involved adaptation at farm level (Smit and Skinner, 2002). Reilly and Schimmelpfennig (1999) suggest evaluating impacts of CC on agriculture by considering vulnerability, where vulnerability in agriculture can be defined in terms of yield, farm profitability, regional economy and hunger. These studies also make clear that lower income populations and marginal agricultural regions, particularly the flood-prone and arid areas, are most vulnerable to CC.

Most adaptations are modifications to existing farming practices and public policy decision-making processes. Therefore, there is a need for a better understanding of the relationship between potential adaptations and existing practices (Smit and Skinner, 2002).

An Adaptation Framework for River Basins

In this research the focus is on adaptation strategies for regional water resources management, which are constrained by the hydro-geographical extensions of a watershed or river basin. The need for integrated basin-wide climate change and water resources studies has been addressed by several studies (Arnell *et al.*, 1996, 2001; Strzepek *et al.*, 1998; Lettenmaier *et al.*, 1999; Kabat and van Schaik, 2003) for a variety of arguments. First, a regional hydrological cycle is bounded by its watershed and is therefore a more appropriate geographical entity than an administrative region or country. Secondly, upstream water-related activities, processes and adaptations have clear effects for downstream water availability. Thirdly, regional water resources management becomes increasingly important in policy making as, for instance, outlined in the EU water framework directive (EU, 2000).

Currently, no specific adaptation framework for river basins exists, although studies point to the relevance of such a framework. Stakhiv (1996) and Frederick (1997) suggest considering the need for adaptation in the water sector, but relevant institutions need to include the process of adaptation in the evaluation criteria that refer to the quality of water resources management. A challenging aspect for developing a generic adaptation framework for river basins is the huge difference in water resources characteristics, environmental controversies and socio-economic issues across river basins. Another issue relates to the above raised question of 'who or what adapts?' For smaller river basins, there is often a water board or a similar institute. For transboundary river basins, however, there is no basin manager or basin-wide institution with a mandate, since country or state borders often determine the jurisdiction of water management. Yet, from a water management perspective, a basin-wide approach still holds for developing and evaluating adaptation strategies, as discussed above.

Thus far, there is a rather *ad hoc* treatment of adaptation in impact assessments and there is too much focus on technical measures (Smit *et al.*, 2000). Moreover, many impact assessment studies within water resources research lack the evaluation adaptations. This can be improved by including stakeholders in the development and evaluation of adaptation, such as outlined in the concept of Integrated Water Resources Management (IWRM).

The above issues are points of departure for developing a generic adaptation framework for river basins called 'AMR' (generic adaptation methodology for river basins). The basic postulates for developing AMR are:

1. That water management in a river basin has a central role in intervening in the water resources system, including adaptation. It is water management, either at basin level or at local level (e.g. a farmer) that implements new adaptations in order to cope with changed climatic conditions (Mendelsohn and Bennett, 1997).
2. The water resources system provides goods and services that are managed so that current and future values are optimized in relation to the objectives of the regional water management policy (Gilbert and Janssen, 1998; EGIS, 2000).
3. Goods and services are expressed as functions of the state of the water resources system, expressed in the terms 'water availability' and 'water quality' (e.g. Gilbert and Janssen, 1998).
4. Climate change is seen as an exogenous influence on the regional water resources system.
5. AMR must allow for evaluating potential adaptation strategies on the basis of a set of decision criteria or indicators that relate to goals and objectives of regional policies (Aerts and Heuvelink, 2002).
6. In order to identify all relevant indicators and to capture the potential adaptations, AMR should allow for active participation of stakeholders in an iterative development and evaluation.
7. AMR preferably builds on existing approaches, such as research by OECD (1993), EEA (1998), Smit *et al.* (1999), Wheaton and MacIver (1999), EGIS (2000) and Barker (2003).

AMR: a goal-based performance framework

The point of departure for developing AMR is to seek a structure that addresses both 'policy objectives' (formulated in regional water management plans, for example) and 'the physical state' of the water resources system. This is done in the following way.

The water resources system can be seen as a productive system that provides 'goods and services' for both humans and ecosystems. These goods and services in a river basin can be broadly classified into four categories: water for food, water for industry, water for nature and water for the human environment (e.g. Gilbert and Janssen, 1998; Kabat and van Schaik, 2003). It is assumed that these goods and services relate directly to the state of the water resources system, which itself can be quantified in both 'water availability' (or quantity) and 'water quality'.

It is the primary task of a water manager optimally to manage the water resources system by securing water quality and allocating water in response to demands for all uses. Hence, the definition of optimal management can be expressed as optimally to use the goods and services by:

- enhancing human welfare;
- enhancing food capacity and security;
- enhancing industrial capacity; and
- enhancing natural ecosystems quality.

These four objectives relate to most water-related management issues in any basin, although basins across the globe obviously differ in water resources characteristics, physical and social environments and therefore in the use of the available goods and services. Every water manager in a given basin has to deal with trade-offs between measures with respect to the four above-mentioned objectives. Each water manager has his or her own priorities in those objectives. For example, within the Rhine Basin in Western Europe, both security against floods and preserving industrial capacity in the form of the number of navigable days have priority above enhancing environmental quality. Water management measures in the Volta Basin, on the other hand, are more targeted to enhance food security through irrigation supply and to preserve hydropower generation.

By quantifying these objectives and priorities, potential measures (including adaptation) can be assessed on their performance. The link to the water resources system determines whether it is feasible to attain such objectives under a given set of adaptation measures. Figure 1.10 schematizes this approach and shows the four objectives of a water manager/policy maker on the left, while the state of the water resources system is presented on the right, expressed in the boxes 'quantity' and 'quality'.

State and decision indicators

In order to operationalize AMR, we need to define sets of indicators that reflect the four aforementioned objectives (*decision indicators*) and that represent the water resources system in detail (*state indicators*). Decision indicators allow for quantifying the performance of water management measures (including adaptation strategies) with respect to the goals (Fig. 1.10).

An indicator has to meet several criteria in order to make it operational: (i) an indicator has to be representative with respect to the goal it represents; (ii) it must be flexible to use and understandable for all stakeholders and users involved in using the framework; (iii) the data needed to measure an indicator must be available; and, finally, (iv) indicators must be generally comparable across the different basins and should preferably aggregate to an index (Cole *et al.*, 1998). Indicators are discussed in detail in Chapters 3 and 4.

The identification of *decision indicators* (DI) may be done using a hierarchical tree, starting with the main objectives and going down to finally arrive at a set of measurable indicators. Figure 1.10 shows an example of such a tree. The four main objectives are presented on the left. From these objectives, a set of intermediate objectives

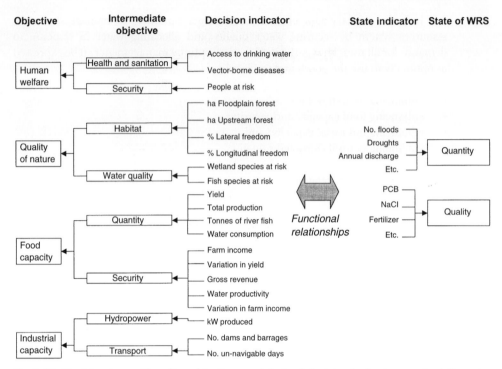

Fig. 1.10. Decision tree with policy objectives and derived decision indicators on the left, and state indicators on the right (Aerts *et al.*, 2003).

can be derived, which together address the intent of the objectives. From the inter-mediate objectives, decision indicators (sometimes called 'criteria') can be derived that allow for quantifying the attainment of water management measures (including adap-tation) with respect to the objectives. Note that objectives have to be quantified as well. Thus the objective of 'enhancing human welfare' is, for example, characterized by: (i) '75% of all inhabitants of a certain river basin have to have access to safe drinking water'; and (ii) 'only 35,000 people are exposed to risks from floods'.

Furthermore, a set of *state indicators* (SI) can be defined that characterizes the state of the water resource system (WRS) of a river basin in terms of water quantity and water quality. These usually relate to quantifiable indicators as 'annual discharge', 'no. of droughts', 'BOD concentration', etc.

EGIS (2000, p. 11) states that 'the functional relations between SIs and DIs are usually expressed in terms of changes'. This is where both simulation models and the role of professionals and stakeholders is required. For example, a hydrological model is capable of simulating the state of the WRS in the form of calculating SIs such as discharges and number of droughts. A subsequent food production model may use these calculations as input for calculating DIs such as 'yield' and 'water productivity'.

Figure 1.10 shows the interlinkage between DIs and SIs. The set of indicators (both DIs and SIs) is derived from the ADAPT Project and obviously not a generic list applicable for all basins. The framework allows inserting new indicators that pertain to a particular case study.

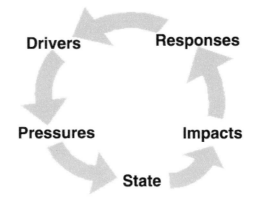

Fig. 1.11. The DPSIR chain of cause–effect relationships (EEA, 1998).

Involving stakeholders

In order to involve stakeholders in the process of developing and evaluating adaptations, we need to provide these stakeholders with an assessment structure. For this, the
DPSIR (Drivers, Pressures, State, Impacts and Response) approach is suggested.
DPSIR allows the structuring of issues and problems in a basin and finally developing responsive adaptation strategies to cope with the impacts of external drivers such
as climate change (OECD, 1993; EEA, 1998). The DPSIR approach was developed
by OECD (1993) and later expanded by the European Environment Agency (EEA,
1998) for indicator-based reporting on the environment. The approach assumes
cause–effect relationships between interacting components of social, economic and
environmental systems (Fig. 1.11).

The DPSIR approach can be divided into five parts, which are here explained
within the context of a water resources system of a river basin:

1. *Driving forces*, such as population growth, economic growth and climate change.
They act upon the
2. *Pressures*, which are activities and/or pollutants resulting from the influence of the
drivers. Most commonly, pressures are very much related to 'issues and problems' of
the water resources system. These pressures cause a change in the
3. *State of water resources system* of a river basin, expressed in terms of (proxi-)
indicators. Through quantifying a change, the
4. *Impacts on the WRS* can be determined. This may induce a
5. *Response* by a water manager in the form of policy measures or technical interventions.

Barker (2003) outlines several weak aspects within the DPSIR approach. The most
important argument in this research is the fact that DPSIR does not address feedback
mechanisms with respect to account for effects of mitigations. For instance, mitigation measures may alter the effect of climate drivers and hence change the way a
response strategy will be implemented.

The weak points of DPSIR are addressed in AMR. First, the effects of mitigation strategies within regional studies can be neglected and will be strictly regarded as exogenous influences. Secondly, all driving forces are explicitly separated from other variables (the DIs and SIs) through treating them as exogenous influences. Thirdly, physical feedback mechanisms will not be considered, except for incorporating CO_2 fertilization (Reilly and Schimmelpfennig, 1999). The framework, in addition, particularly addresses the involvement of stakeholders in an iterative process, and hence allows for policy feedbacks within the cause–effect chain. Finally, the integration of DPSIR and the goal-based framework allows for an evaluation of adaptation strategies through using the decision indicators as evaluation criteria. Its simplicity, though, is a strong point and DPSIR strongly 'steers' the response strategy towards the problem.

AMR in practice

Figure 1.2 shows the total AMR framework. The circle shows the participative approach where stakeholders in a river basin play a key role. They are confronted with: (i) exogenous influences such as population growth and climate change; (ii) how these impact the WRS; and (iii) how these impacts affect the goals and decision variables. Next, the framework allows for: (iv) deciding which potential adaptation strategies can be applied to respond to the impacts. And finally, (v) each of these strategies can be evaluated by measuring their performance against the pre-set goals.

The following example explains the use of AMR by going through the indicated boxes in Fig. 1.2. Consider a river basin that is potentially vulnerable to climate change. (1) Scenarios point to an increase in temperature and hence both increased evaporation and droughts in the basin. (2) Hydrological models have calculated the water availability according to both a Business As Usual (BAU) scenario and a CC scenario. (3) Next, food and wetland models use these water availability figures in order to calculate the effects of CC for two decision indicators: 'Rice production' in tonnes per hectare and 'Preserved wetlands' in hectares. First, an increase in rice production under CC (from 2000 to 2500 t/ha) is projected, but is much less than would be expected without CC (the target is 3500 t/ha). Secondly, the area of viable wetlands will decrease under CC from 150,000 to 100,000 ha, while the national wetland protection plan aims for a stable area of 150,000 ha of wetland in the future. (4) Stakeholders, including water managers, have selected two possible adaptations. One is targeted to alleviate impacts on rice production in the basin by increasing irrigation efficiency (A1, Table 1.1). The other adaptation measure (A2) aims at preserving downstream wetlands at the expense of irrigated rice production. (5) The previous food and wetlands models have calculated the effects of implementing these adaptations (Table 1.1). It appears in this example that A2 performs better than A1 when considering both indicators as equally important. Although A1 better achieves the goal formulated for 'rice production', it shows poorer performance on achieving 'wetland preservation'.

Note that, in practice, decision indicators are certainly not preferred equally and they have to be prioritized in order to finally use them in an evaluation of different management options. This process of prioritizing and evaluating using indicators (also referred to as 'criteria') is called 'Multi Criteria Analysis'.

Table 1.1. Example of measuring the performance of adaptation strategies (A1 and A2) against a set of two decision indicators (rice production and wetland preservation).

	Current situation	Objective	Future with CC		Adaptation			
					Future with CC + A1		Future with CC + A2	
Decision indicators				Slack 1: % from goal		Slack 2: % from goal		Slack 3: % from goal
Rice production (t/ha)	2,000	3,500	2,500	28	3,000	14	2,800	20
Wetland preserved (ha)	150,000	150,000	100,000	33	80,000	47	120,000	20

In the previous example, the adaptation strategy A1 'increasing irrigation efficiency' is probably a set of individual measures. For instance, increasing irrigation efficiency can be achieved through better irrigation techniques, other crops, or new crop rotation schemes. Thus, an adaptation strategy is seen as a set of individual measures that through joint implementation can achieve the different objectives. We have seen in the same example that these are often conflicting objectives, as adaptation strategy A2 is favourable for preserving nature, while A1 is primarily targeted at food security. On many occasions, though, a water manager seeks to satisfy all objectives and hence must combine a variety of measures, which together form a sustainable and balanced adaptation strategy.

Concluding Remarks

In this chapter we have provided an extensive overview of planning for adaptation for water managers, with a special focus on climate change. Although some references have been provided, adaptation strategies are still in their early stages in comparison with other climate change related issues, such as mitigation and projection. We have provided a systematic approach, here referred to as AMR: Adaptation Methodology for River Basins. AMR aims to link policy objectives with the physical state of water resources in a river basin. The four focal areas are: water for drinking, water for food, water for industry and water for ecosystems, while the sets of 'decision indicators' and 'state indicators' complete the entire framework.

In the subsequent chapters of this book, and especially the seven chapters referring to individual basins, this approach will be demonstrated. Not every chapter will exactly copy the approach described here, but the key components can be found in each basin chapter. Readers may wish to pay special attention to the way the four focal areas of AMR (domestic, industry, food, ecosystems) are addressed in the different river basins.

References

Aerts, J.C.J.H. and Heuvelink, G.B.M. (2002) Using simulated annealing for resource allocation. *International Journal of Geographical Information Science* 16, 571–587.

Aerts, J.C.J.H., Lasage, R. and Droogers, P. (2003) *A Framework for Evaluating Adaptation Strategies.* Institute for Environmental Studies report R-03/08. Vrije Universiteit Amsterdam, The Netherlands.

Alcamo, J. and Heinrichs, T. (2001) Critical regions for water and vulnerability. *Proceedings Global Change Open Science Conference 'Challenges of a Changing Earth',* 10–13 July 2001, Amsterdam, The Netherlands.

Allen, M.R. and Ingram, W.J. (2002) Constraints on future changes in climate and the hydrologic cycle. *Nature* 419, 224–232.

Arnell, N.W., Bates, B., Lang, H., Magnuson, J.J., Mulholland, P., Fisher, S., Liu, C., McKnight, D., Starosolszky, O., Taylor, M., Aquize, E., Arnott, S., Brakke, D., Braun, L., Chalise, S., Chen, C., Folt, C.L., Gafny, S., Hanaki, K., Hecky, R., Leavesly, G.H., Lins, H., Nemec, J., Ramasastri, K.S., Somlyódy, L. and Stakhiv, E. (1996) Hydrology and freshwater ecology. In: Watson, R.T., Zinyowera, M.C. and Moss, R.H. (eds) *Climate Change 1995 – Impacts, Adaptations and Mitigation of Climate Change: Scientific-Technical Analyses.* Contribution of Working Group II to the Second Assessment of the Intergovernmental Panel on Climate Change. Cambridge University Press, Cambridge, pp. 325–363.

Arnell, N.W.C., Liu, R., Compagnucci, L., Da Cunha, C., Howe, K., Hanaki, G., Mailu, I., Shiklomanov, A., Döll, P., Becker, A. and Zhang, J. (2001) Hydrology and water resources. In: McCarthy, J.J., Canziani, O., Leary, N.A., Dokken, D.J. and White, K.S. (eds) *Climate Change 2001 – Impacts, Adaptation and Vulnerability. Contribution of Working Group II to the Third Assessment Report of the Intergovernmental Panel on Climate Change.* Cambridge University Press, Cambridge, pp. 191–233.

Barker, T. (2003) Representing global climate change, adaptation and mitigation. *Global Environmental Change* 13, 1–6.

Bazzaz, F. and Sombroek, W. (1996) *Global Change and Agricultural Production.* Food and Agriculture Organization of the United Nations (FAO), John Wiley & Sons, New York.

Burton, I., Smith, J. and Lenhart, S. (1998) Adaptation to climate change: theory and assessment. In: Feenstra, J.F., Burton, I., Smith, J.B. and Tol, R.S.J. (eds) *UNEP Handbook on Methods for Climate Change Impact Assessment and Adaptation Strategies.* Institute for Environmental Studies, Vrije Universiteit, Amsterdam.

Cole, D.C., Eyles, J. and Gibson, B.L. (1998) Indicators of human health in ecosystems: what do we measure? *The Science of the Total Environment* 224, 201–213.

Cosgrove, W.J. and Rijsberman, F.R. (2000) *World Water Vision – Making Water Everybody's Business.* Earthscan Publications, London.

Dvorák, V., Hladny, J. and Kaspárek, L. (1997) Climate change hydrology and water resources impact and adaptation for selected river basins in the Czech Republic. *Climatic Change* 36, 93–106.

EEA (1998) *Towards Environmental Pressure Indicators for the EU – First Edition.* European Environment Agency, Copenhagen.

EGIS (2000) *Blue Accounting – Introduction to a Methodology for Monitoring and Assessing the Functionality of the Water Resources System.* EGIS technical note 15. Dhaka, Bangladesh.

EU (2000) Directive 2000/60/EC, EU Water Framework Directive. Brussels, Belgium.

FAO (2003) *World Agriculture: Towards 2015/2030. An FAO Perspective.* Bruinsma, J. (ed.) Earthscan Publications, London.

Fischer, G., Shah, M., van Velthuizen, H. and Nachtergaele, F.O. (2001) *Executive Summary Report: Global Agro-ecological Assessment for Agriculture in the 21st Century.* International Institute for Applied Systems Analysis, Laxenburg.

Frederick, K.D. (1997) Adapting to climate impacts on the supply and demand for water. *Climatic Change* 37, 141–156.

Gilbert, A.J. and Janssen, R. (1998) Use of environmental functions to communicate the values of a mangrove ecosystem under different management regimes. *Ecological Economics* 25, 323–346.

IPCC (1998) *The Regional Impacts of Climate Change – An Assessment of Vulnerability.* Watson, R.T.,

Zinyowera, M.C. and Moss R.H. (eds) Cambridge University Press, Cambridge.

IPCC (2001a) *Climate Change 2001 – The Scientific Basis. Contribution of Working Group I to the Third Assessment Report of the Intergovernmental Panel on Climate Change.* Houghton, J.T., Ding, Y., Griggs, D.J., Noguer, M., van der Linden, P.J., Dai, X., Maskell, K. and Johnson, C.A. (eds) Cambridge University Press, Cambridge.

IPCC (2001b) *Climate Change 2001 – Impacts, Adaptation and Vulnerability. Contribution of Working Group II to the Third Assessment Report of the Intergovernmental Panel on Climate Change.* McCarthy, J.J., Canziani, O.F., Leary, N.A., Dokken, D.J. and White, K.S. (eds) Cambridge University Press, Cambridge.

Kabat, P. and van Schaik, H. (2003) *Climate Changes the Water Rules: How Water Managers Can Cope with Today's Climate Variability and Tomorrow's Climate Change.* Dialogue on Water and Climate. Printfine, Liverpool.

Kane, S. and Shogren, J. (2000) Linking adaptation and mitigation in climate change policy. *Climatic Change* 45, 75–101.

Kelly, P.M. and Adger, W.N. (2000) Theory and practice in assessing vulnerability to climate change and facilitating adaptation. *Climatic Change* 47, 325–352.

Klein, R.J.T. and Tol, R.S.J. (1997) *Adaptation to Climate Change: Options and Technologies.* An overview paper. Technical Paper FCCC/TP/1997/3, United Nations Framework Convention on Climate Change Secretariat, Bonn, Germany.

Klein, R.J.T. and MacIver, D.C. (1999) Adaptation to climate variability and change: methodological issues. *Mitigation and Adaptation Strategies for Global Change* 4, 189–198.

Leary, N.A. (1999) A framework for benefit–cost analysis of adaptation to climate change and climate variability. *Mitigation and Adaptation Strategies for Global Change* 4, 307–318.

Lettenmaier, D.P., Wood, A.W., Palmer, R.N., Wood, E.F. and Stakhiv, E.Z. (1999) Water resources implications of global warming: a U.S. regional perspective. *Climatic Change* 43, 537–579.

MacIver, D.C. and Dallmeier, F. (2000) IPCC Workshop on adaptation to climate variability and change: adaptive management. *Journal of Environmental Monitoring and Assessment* 61, 1–8.

Mendelsohn, R. and Bennett, L.L. (1997) Global warming and water management: water allocation and project evaluation. *Climatic Change* 37, 271–290.

Middelkoop, H., Daamen, K., Gellens, D., Grabs, B., Kwadijk, J.C.J., Lang, H., Parmet, B.W.A.H., Schädler, B., Schulla, J. and Wilke, K. (2001) Impact of climate change on hydrological regimes and water resources management in the Rhine basin. *Climatic Change* 49, 105–128.

Mirza, M.M.Q. (2002) Global warming and changes in the probability of occurrence of floods in Bangladesh and implications. *Global Environmental Change* 12, 127–138.

OECD (1993) *OECD Core Set of Indicators for Environmental Performance Reviews.* Environment Monograph No. 83. Paris, France.

Parry, M., Rosenzweig, C., Iglesias, A., Fischer, G. and Livermore, M. (1999) Climate change and world food security: a new assessment. *Global Environmental Change* 9, 51–67.

Pielke, R.A. Jr (1998) Rethinking the role of adaptation in climate policy. *Global Environmental Change* 8, 159–170.

Pittock, A.B. and Jones, R.N. (2000) Adaptation to what and why? *Environmental Monitoring and Assessment* 61, 9–35.

Reilly, J.M. and Schimmelpfennig, D. (1999) Agricultural impact assessment, vulnerability, and the scope for adaptation. *Climatic Change* 43, 745–788.

Reilly, J. and Schimmelpfennig, D. (2000) Irreversibility, uncertainty, and learning: portraits of adaptation to long-term climate change. *Climatic Change* 45, 253–278.

Smakhtin, V., Revenga, C. and Döll, P. (2004) Putting the water requirements of freshwater ecosystems into the global picture of water resources assessment. *Environmental Conservation* (in press).

Smit, B. (1993) *Adaptation to Climatic Variability and Change: Report of the Task Force on Climate Adaptation.* Environment Canada, Guelph.

Smit, B. and Skinner, M. (2002) Adaptation options in agriculture to climate change: a typology. *Mitigation and Adaptation Strategies for Global Change* 7, 85–114.

Smit, B., Burton, I., Klein, R. and Street, R. (1999) The science of adaptation: a framework for assessment. *Mitigation and Adaptation Strategies for Global Change* 4, 199–213.

Smit, B., Burton, I., Klein, R. and Wandel, J. (2000) An anatomy of adaptation to climate change and variability. *Climatic Change* 45, 223–251.

Smit, B., Pilifosova, O., Burton, I., Challenger, B., Huq, S., Klein, R.J.T., Yohe, G., Adger, N., Downing, T., Harvey, E., Kane, S., Parry, M., Skinner, M., Smith, J. and Wandel, J. (2001) Adaptation to climate change in the context of sustainable development and equity. In: McCarthy, J.J., Canziani, O., Leary, N.A., Dokken, D.J. and White, K.S. (eds) *Climate Change 2001 – Impacts, Adaptation and Vulnerability. Contribution of Working Group II to the Third Assessment Report of the Intergovernmental Panel on Climate Change.* Cambridge University Press, Cambridge, pp. 877–912.

Stakhiv, E.Z. (1996) Managing water resources for climate change adaptation. In: Smith, J., Bhatti, N., Menzhulin, G., Benioff, R., Budyko, M.I., Campos, M., Jallow, B. and Rijsberman, F. (eds) *Adapting to Climate Change: An International Perspective.* Springer, New York, pp. 243–264.

Strzepek, K.M., Campos, M., Kaczmarek, Z., Sanchez, A., Yates, D., Carmichael, J., Chavula, G.M.S., Chirwa, A.B., Mirza, M., Minarik, B. and Warrick, R. (1998) Water resources. In: Feenstra, J.F., Burton, I., Smith, J.B. and Tol, R.S.J. (eds) *UNEP Handbook on Methods for Climate Change Impact Assessment and Adaptation Strategies.* Institute for Environmental Studies, Vrije Universiteit, Amsterdam.

Tol, R.S.J. (1996) A systems view of weather disasters. In: Downing, T.E., Olsthoorn, A.A. and Tol, R.S.J. (eds) *Climate Change and Extreme Events: Altered Risk, Socio-economic Impacts and Policy Responses.* Institute for Environmental Studies report R-96/04, Vrije Universiteit, Amsterdam, pp. 17–34.

Tol, R.S.J., Fankhauser, S. and Smith, J.B. (1998) The scope for adaptation to climate change: what can we learn from the impact literature? *Global Environmental Change* 8, 109–123.

UNFCCC (1992) *United Nations Framework Convention on Climate Change.* United Nations, New York. Available at: http://unfccc.int/resource/ccsites/senegal/conven.htm#art4

UNFCCC (1997) *Kyoto Protocol to the United Nations Framework Convention on Climate Change.* FCCC/CP/L7/ Add.1, 10 December. United Nations, New York. Available at: http://unfccc.int/resource/ccsites/senegal/conven.htm#art4

Wheaton, E.E. and MacIver, D.C. (1999) A framework and key questions for adapting to climate variability and change. *Mitigation and Adaptation Strategies* 4, 215–225.

Yohe, G. and Dowlatabadi, H. (1999) Risk and uncertainties, analysis and evaluation: lessons for adaptation and integration. *Mitigation and Adaptation Strategies for Global Change* 4, 319–329.

Zeidler, R.B. (1997) Climate change vulnerability and response strategies for the coastal zone of Poland. *Climatic Change* 36, 151–173.

2 Evaluating Downscaling Methods for Preparing Global Circulation Model (GCM) Data for Hydrological Impact Modelling

LAURENS M. BOUWER,[1] JEROEN C.J.H. AERTS,[1] GUIDO M. VAN DE COTERLET,[1] NICK VAN DE GIESEN,[2] AMBRO GIESKE[3] AND CHRIS MANNAERTS[3]

[1]Institute for Environmental Studies (IVM), Faculty of Earth and Life Sciences, Vrije Universiteit Amsterdam, The Netherlands; [2]Centre for Development Research (ZEF), Bonn University, Germany; [3]International Institute for Geo-information Science and Earth Observation (ITC), Enschede, The Netherlands

Introduction

For climate impact assessments, different scenarios can be constructed. The main types of scenarios are synthetic scenarios, analogue scenarios and scenarios from so-called global circulation models (GCMs). According to the Intergovernmental Panel on Climate Change (IPCC), GCMs offer the most credible tools for estimating the future responses to increased greenhouse gas emissions (IPCC-TGCIA, 1999). Although most models largely agree on the expected large-scale pattern of climate change, there are still important uncertainties in the regional projections.

The most widely used climate change projections are provided by the IPCC and are described in their Third Assessment Report (IPCC, 2001). These projections are based on outputs of GCMs and are used in many studies that assess impacts of climate change on hydrology. There are many difficulties in using the GCM, however. Arnell et al. (1996) identify two major weaknesses: (i) GCMs are poorly capable of coupling land-surface and atmospheric hydrological cycles, particularly simulating regional precipitation and climate extremes; and (ii) the spatial and temporal resolution of current climate model outputs is too coarse to be used directly in (regional) hydrological models. This chapter focuses on handling the second issue.

GCM outputs cover the entire globe and generally have coarse spatial resolutions, with grid-cell size typically over $2.5° \times 2.5°$ (approximately 250×250 km^2). Hydrological effects and adaptation to changes in climate in river basins, however, are mainly taking place at local to regional scales that lie within just one or a few GCM grid-cells. Furthermore, GCM runs for historical time slices tend to show local and regional discrepancies with respect to measured variables such as temperature and precipitation. Especially with respect to precipitation, a central input for hydrological models used in the ADAPT project, these discrepancies are too large to be overlooked and need to be compensated for.

For correcting GCM outputs at the regional level, two basic approaches exist, which are referred to as statistical and dynamical downscaling. For regional impact studies, two basic downscaling steps are needed before climate model outputs can be used meaningfully (McGuffie and Henderson-Sellers, 1997). The first step is the spatial downscaling of the coarse GCM data to a level that reflects actual gradients and differences within the region. The second step is to transform the output in such a way that the main statistical properties of historically observed data match those of the transformed climate model output. This second step is the statistical downscaling. A second approach exists for using GCM data in regional studies and is called dynamical downscaling. A dynamical approach uses the output of GCMs as boundary conditions for the region of interest. A fully physical weather model is then used to calculate future climate at a regional scale. An important advantage of dynamical models is that they account for local conditions, which may include changes in land cover. The overriding disadvantage is that such dynamic models need tremendous computing resources.

Research has shown that certain downscaling techniques can be used to prepare global GCM outputs for regional hydrological modelling. Although there have been many developments in this area (for an overview, see for example, Wilby and Wigley, 1997), the major problems described above still exist (Arnell *et al.*, 2001). Since the ADAPT project focused on developing and evaluating adaptation strategies for river basins, regional climate projections were a prerequisite for conducting this study. Knowing the above-mentioned drawbacks of using GCM outputs in regional studies, the aim of this chapter is to show a selected set of relatively simple downscaling methods and to assess the differences in applying these in regional studies in terms of uncertainty. The main goals of this chapter are to:

- describe four downscaling methods used in the ADAPT project;
- apply the four downscaling techniques to precipitation and temperature projections according to two GCM models: HadCM3 and ECHAM4, using two emission scenarios; and
- statistically compare the downscaling results as to how they fit the measured data.

Available Data

Observed baseline climate data

Downscaling techniques are applied so that the GCM data fit measured data within a baseline period. The reference or baseline period was chosen between 1961 and 1990, as suggested by the World Meteorological Organisation (WMO).

As observed data in this period, the CRU TS 1.2 dataset provided by the Climatic Research Unit (CRU) was used (New *et al.*, 2000). This dataset provides interpolated monthly precipitation and temperature values for global land surfaces based on observations between 1901 and 1998 on a 0.5° × 0.5° grid. Precipitation is expressed in millimetres and average temperature in 10ths of degrees Celsius. These datasets are not corrected for ambient factors such as urban development or land use change and thus cannot be used to investigate actual climate change. For example, according to New *et al.* (2000), the CRU dataset is on average about 0.1°C warmer in the Northern Hemisphere than the corrected dataset that was used by Jones (1994) to investigate global temperature change.

Climate change scenarios

GCM simulations are calculated using so-called forcing scenarios. These scenarios are based on projected socio-economic developments over periods of time and the related emission of greenhouse gases (GHG) and aerosols. The most recent emission scenarios are reported in the IPCC Special Report on Emission Scenarios (SRES; IPCC, 2000).

The SRES report uses the following terminology.

- Storyline: a narrative description of a scenario (or a family of scenarios), highlighting the main scenario characteristics and dynamics, and the relationships between key driving forces.
- Scenario: projections of a potential future, based on a clear logic and a quantified storyline.
- Scenario family: one or more scenarios that have the same demographic, politico-societal, economic and technological storylines.

There are six scenario groups that span a wide range of uncertainty. These encompass four combinations of demographic change, social and economic development, and broad technological developments, corresponding to four families (A1, A2, B1, B2), each with an illustrative 'marker' scenario (Fig. 2.1). The scenarios in the SRES report do not include additional climate initiatives, which means that no scenarios are included that explicitly assume implementation of the United Nations Framework Convention for Climate Change (UNFCCC) or the emissions targets of the Kyoto Protocol. However, GHG emissions are directly affected by non-climate change policies designed for a wide range of other purposes. Furthermore, government policies can, to varying degrees, influence the GHG emission drivers such as demographic change, social and economic development, technological change, resource use and pollution management. This influence is broadly reflected in the storylines and resulting scenarios.

The six marker scenarios that were developed can be briefly described as (Parry, 2002):

- A1FI: A future world of very rapid economic growth, and intensive use of fossil fuels.
- A1T: A future world of very rapid economic growth, and rapid introduction of new and more efficient technology.

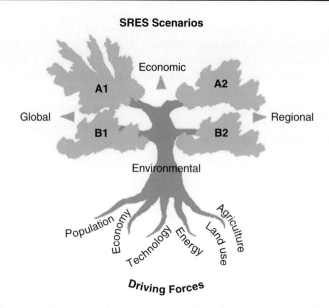

Fig. 2.1. Schematic illustration of SRES scenarios. The schematic diagram illustrates that the scenarios build on the main driving forces of greenhouse gas emissions. Each scenario family is based on a common specification of some of the main driving forces (IPCC, 2000).

- A1B: A future world of very rapid economic growth, and a mix of technological developments and fossil fuel use.
- A2: A future world of moderate economic growth, more heterogeneously distributed and with a higher population growth rate than in A1.
- B1: A convergent world with rapid change in economic structures, 'dematerialization', introduction of clean technologies, and the lowest rate of population growth.
- B2: A world in which the emphasis is on local solutions to economic, social and environmental sustainability, intermediate levels of economic development and a lower population growth rate than A2.

The IPCC has a number of recommendations for the application of climate change scenarios from global circulation models (GCM) data (IPCC-TGCIA, 1999). These recommendations address the relatively coarse resolution of GCM outputs compared to required resolutions in regional studies, the validity of the simulated historical climate and the representativeness of these scenarios for a range of climatic changes.

The IPCC (2000) report on scenarios recommends the use of more than one scenario family in analyses. Furthermore, the IPCC Task Group on Scenarios for Climate Impact Assessment (TGCIA) has asked climate-modelling centres to give priority to the A2 and B2 scenarios when running GCMs (Parry, 2002). At the time of this research only the A2 and B2 scenarios were available. The A2 scenario represents a relatively 'high' climate change scenario with CO_2 concentrations reaching 850 parts per million (ppm) by 2100, compared to about 360 ppm now, while B2 represents a relatively 'low' climate change scenario with CO_2 concentrations reaching 600 ppm by 2100. Global average temperatures would increase by approximately 3.8 and 2.6°C, respectively, relative to 1990.

The main uncertainties within the SRES emission scenarios originate from (Allen *et al.*, 2001):

1.   The difficulty of assigning probabilities of socio-economic trends and resulting emissions.
2.   The difficulty of obtaining consensus ranges for quantities like climate sensitivity.
3.   The possibility of a non-linear response in the carbon cycle or ocean circulation.

For regional studies it can be noted that a fourth source of uncertainty exists, which is the difference in estimates of different models of regional climate change for the same mean global warming (see also Chapter 1). Wigley and Raper (2001), in their global probabilistic analysis, give a 90% probability interval for global temperatures being between 1.7 and 4.9°C higher by 2100.

Global circulation model output

The IPCC Data Distribution Centre (IPCC-DDC, http://ipcc-ddc.cru.uea.ac.uk/) provides online access to outputs of eight GCMs. Two GCMs were chosen that have been widely used for various climate impact assessments. These are the Hadley Climate Model 3 (HadCM3) from the Hadley Centre for Climate Prediction and Research in Bracknell, UK and the European Climate Model 4 with the OPYC3 ocean circulation model (ECHAM4/OPYC3) from the Max Planck Institute für Meteorologie in Hamburg, Germany, from here on called ECHAM4.

The HadCM3 model output is an ensemble mean of three GCM runs using the A2 scenario as input, and of two runs using the B2 scenario. The HadCM3 data are available on a 2.5° × 3.75° grid, while the ECHAM4 data are available on a 2.8125° × 2.8125° grid. Both models provide data for the period 1950–2099 and 1990–2100, respectively. It was decided to apply the downscaling techniques to the whole output period; for further analysis in the basins, only two future time slices were selected, one in the near future or medium term (2010–2039) and one in the more distant future or long term (2070–2099) (see the river basins in Chapters 5–11). As stated above, data for the baseline period 1961–1990 were used in order to first assess the GCM performance without downscaling and next measuring performance after applying downscaling. Unfortunately, the ECHAM4 SRES runs only include data for the period 1990–2100.

Downscaling Methods

In this section, one spatial and four statistical downscaling techniques are explained and applied to a single set of GCM output data from the HadCM3 model. The GCM was forced with greenhouse gas emissions according to the IPCC SRES A2 scenario (IPCC, 2000). These data sets were retrieved from the IPCC-DDC website. First,

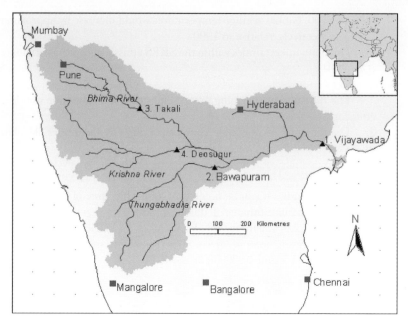

Fig. 2.2. Location of the Krishna River Basin in central India.

Kriging interpolation was applied to spatially resample the coarse GCM data to the CRU resolution. Next, four different statistical downscaling methods of increasing complexity were applied, in order to assess the relative benefit of more complex techniques.

All downscaling techniques were tested on data for the Krishna River Basin in India. Although this area was not included in the ADAPT project, enough information and data were available to let it serve as a test case. For the analysis of the Krishna River Basin, we used the most recent version of the CRU dataset (TS 2.0; Mitchell, T.D., Carter, T.R., Jones, P.D., Hulme, M. and New, M., 2003, unpublished). The same spatial techniques and the best statistical downscaling technique were finally applied to all ADAPT basins.

The Krishna River Basin is the second largest river in peninsular India and stretches over an area of 258,948 km^2. The basin represents almost 8% of the surface area of India and in 1991 was inhabited by 60.8 million people. Major tributaries include the Bhima River in the north and the Tungabhadra River in the south (see Fig. 2.2). The climate is characterized by sub-tropical conditions, with considerable rainfall in the mountains of the Western Ghatts and much drier conditions in the basin interior. The river terminates at the Krishna delta in the Bay of Bengal.

Before the GCM datasets could be used, some simple conversion had to be applied in order to match the unit of the data to the unit that is used in the observed CRU baseline data. The original GCM data on precipitation are in millimetres per day and were converted to millimetres per month, as in the CRU baseline data. The temperature values of the GCM data were first changed from Kelvin to tens of degrees Celsius using Equation (1):

$$t' = 10 \times (t - 273.15) \tag{1}$$

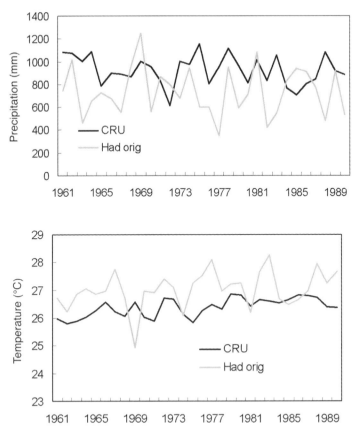

Fig. 2.3. Comparison of annual observed (CRU) and original modelled data (Had orig). Precipitation is shown at the top and temperature at the bottom, both for the baseline period (1961–1990).

where t' is the adjusted temperature in tens of degrees Celsius and t is the original value in Kelvin.

Figure 2.3 shows the comparison of the raw GCM data (without applying any downscaling technique) against the observed CRU data for the Krishna River Basin for the baseline period. Note that individual years do not match, because a GCM is not able to simulate individual years accurately. More important is the comparison between simulated and observed average values and the variability. Although the average precipitation and temperature do more or less match, the variability of the model appears too high for both parameters.

Spatial downscaling

Figure 2.4 schematically illustrates the difference between the CRU and GCM (HadCM3) grid boxes. It shows a simple example where coarse GCM cells (left) must be interpolated towards a finer CRU grid (right). From this the problem becomes

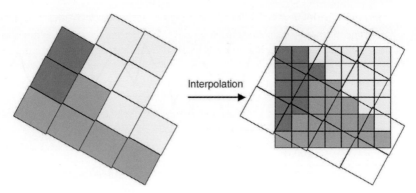

Fig. 2.4. Simple representation of applying interpolation on a coarse GCM grid (left) to arrive at a finer CRU grid (right). Note that both grids have different geographic projections in this example.

clear: when studying a site that is just overlapping with (parts of) just a few model grid-boxes, the simulated GCM data are likely not to be representative for the specific location. Hence firstly, the relatively coarse GCM data must be converted to the finer CRU resolution of $0.5° \times 0.5°$. This spatial downscaling process uses an interpolation technique to derive cell values in those locations where GCM data are missing. The interpolation method used here is called Kriging, which is a standard procedure in a geographic information system (GIS). For details see Burough (1986).

Statistical downscaling

Statistical downscaling approaches use both GCM and measured data for the same period to adjust the GCM data such that certain statistical properties (mean, standard deviation, range, correlation matrices, etc.) of the adjusted GCM data are similar to the same properties of the measured dataset. The advantage of this method is its relative simplicity, making it often the only feasible option available.

Different statistical downscaling methods exist that ensure that GCM and historical data have similar statistical properties. In this section, four methods will be explained and evaluated.

Method 1: Corrections based on annual averages

The first statistical transformation method is a simple approach based on Alcamo *et al.* (1997). The method uses a correction based on the difference between average annual observations and the average annual modelled GCM data. For precipitation a factor is used, while for temperature an absolute difference is used. Corrected GCM temperatures are calculated for each month i using Equation (2):

$$t'_{GCM} = t_{GCM} + (\bar{t}_{GCM} - \bar{t}_{CRU}) \tag{2}$$

where t'_{GCM} is the corrected temperature for a particular month, t_{GCM} the original simulated temperature for a particular month, $\bar{t}_{GCM}$ the simulated average annual temperature from the GCM for the 30-year baseline period, and $\bar{t}_{CRU}$ the observed average annual temperature from the GCM over the 30-year baseline period.

For precipitation, the following correction can be applied (Equation (3)):

$$p'_{GCM} = p_{GCM} \times (\bar{p}_{GCM} / \bar{p}_{CRU}) \tag{3}$$

where p'_{GCM} is the corrected precipitation for a particular month, p_{GCM} the original simulated precipitation for a particular month, $\bar{p}_{GCM}$ the simulated average annual precipitation from the GCM over the 30-year baseline period, and $\bar{p}_{CRU}$ the observed average annual precipitation over the 30-year baseline period.

Method 2: Corrections based on monthly averages

This method is similar to method 1, but it uses the difference between the individual monthly averages (January–December) of observations and modelled GCM data.

Method 3: Corrections based on monthly averages, spatially explicit

This method uses a monthly correction factor as in method 2, but this factor is calculated for each individual pixel. Hence, this method spatially differentiates corrections.

Method 4: Corrections using the standard deviation

The fourth method is a method in which the GCM data are not only corrected against the average observed climate but also for the observed variance. This is done by inserting the standard deviation in the formula. The formula is constructed in such a way that both the average climate and the variability of simulated series after correction exactly match the observation (Equation (4)):

$$a'_{GCM} = \left(\frac{a_{GCM} - \bar{a}_{GCM,j}}{\sigma_{GCM,j}} \right) \times \sigma_{CRU,j} + \bar{a}_{CRU,j} \tag{4}$$

where a'_{GCM} is the corrected climate parameter (total precipitation or average temperature) in a particular month, a_{GCM} the un-corrected simulated climate parameter, $\bar{a}_{GCM,j}$ the average simulated climate parameter in the corresponding month j (January–December) over the 30-year baseline period, $\sigma_{GCM,j}$ the standard deviation of the simulated climate parameter over the baseline period, $\sigma_{CRU,j}$ the standard deviation of the observed climate parameter over the baseline period, and $\bar{a}_{CRU,j}$ the average observed climate parameter over the baseline period.

Evaluation of Statistical Downscaling Methods

Comparison of the observed and simulated climate

The statistical differences can be assessed between the modelled (statistically transformed) and the observed CRU datasets. Firstly, spatial downscaling was applied and the discrepancies between the observed (CRU) and the downscaled HadCM3 monthly are illustrated in Fig. 2.5. In this figure, frequencies and cumulative frequencies for

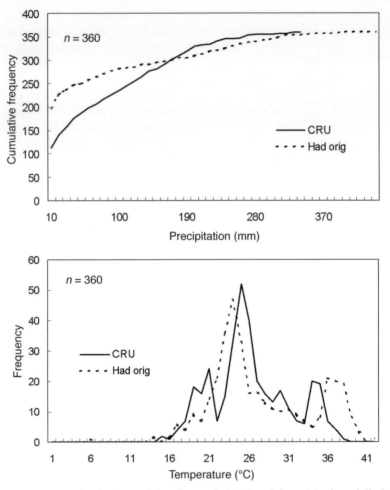

Fig. 2.5. Frequency distributions of the observed (CRU) and the original modelled HadCM3 data of monthly precipitation (top) and temperature (bottom) for the Krishna River Basin over the baseline period (1961–1990).

temperature and monthly precipitation are plotted for both datasets. The figure shows that for both temperature and precipitation rather large discrepancies exist, especially for precipitation.

Application of the statistical downscaling methods

Next, after the spatial downscaling of the data, all four statistical downscaling methods were applied. The same frequency curves as in Fig. 2.5 were drawn. Figure 2.6 shows the frequency distribution and the cumulative frequency of the modelled data for the baseline period (1961–1990) following the downscaling methods described earlier. The figure shows that the spatial method (method 3) and especially

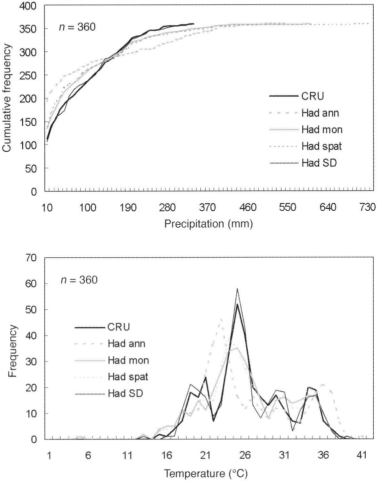

Fig. 2.6. Frequency distributions of the simulated monthly precipitation (top) and temperature (bottom) for the Krishna River Basin over the baseline period (1961–1990), corrected using four different methods. Also shown are the observed (CRU) data.

the standard deviation method (method 4) match the measured data best. For example, it can be seen from the frequency curves for temperature that the peaks resulting from applying the annual method (method 1) and the monthly method (method 2) are still quite different, whereas the other methods adjust the GCM data to match the CRU data.

Figure 2.7 shows the root mean square error (RMS) of the original and corrected model precipitation (left) and temperature (right) data, as compared to the observed (CRU) data for the baseline period (1961–1990). It can be seen that methods 3 and 4 perform best when comparing RMS values against the RMS value of the original HadCM3 data. After applying the annual method (method 1) to precipitation data the RMS error is even larger than in the original HadCM3 data.

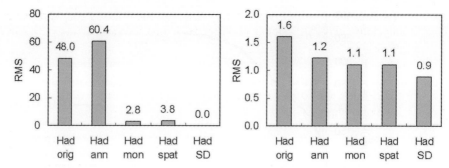

Fig. 2.7. Root mean square error (RMS) of the original and corrected precipitation (left, mm) and temperature (right, °C) model data, as compared to the observed (CRU) data for the baseline period (1961–1990).

For completeness, Table 2.1 provides an overview of some statistical parameters in the performance of the four downscaling techniques. From this table it can be seen that indeed all four methods correct the mean value of the HadCM3 dataset to a value more close to the CRU dataset. However, large differences still exist (especially for methods 1 and 2) in terms of correcting the data for extreme values. For instance, the 10th and 90th percentile values are still different for methods 1 and 2 when comparing these to the CRU data.

After the application of the methods to the baseline period, it is also possible to extend the methods to the future time slice (A2 scenario) for the Krishna River Basin, over the period 2070–2099. The results of this are shown in Table 2.2.

Table 2.1. Comparison of the mean, minimum, maximum, standard deviation, 10th and 90th percentiles of the observed (CRU) and modelled (Had) monthly climate data for the baseline period (1961–1990).

	CRU	Had original	Had annual	Had monthly	Had spatial	Had SD
Precipitation						
Mean	77.2	61.5	77.2	76.2	76.4	77.2
Median	42.8	6.2	7.8	30.0	25.0	56.2
Minimum	0.3	0.0	0.0	0.0	0.0	0.9
Maximum	335.7	436.5	547.3	592.9	730.4	339.5
SD	81.7	98.7	123.8	102.0	108.4	81.3
10th	1.0	0.4	0.5	0.5	0.5	1.8
90th	195.8	237.2	297.4	215.8	223.9	199.4
Temperature						
Mean	26.4	27.0	26.4	26.4	26.4	26.4
Median	26.0	26.0	25.4	26.1	25.9	26.0
Minimum	20.9	16.3	15.7	15.6	15.6	19.8
Maximum	32.0	34.2	33.6	32.7	32.7	32.9
SD	2.5	3.1	3.1	2.7	2.7	2.5
10th	23.1	23.6	23.0	22.9	22.9	23.0
90th	30.4	31.9	31.3	30.4	30.4	30.4

Table 2.2. Comparison of the mean, minimum, maximum, standard deviation, 10th and 90th percentiles for the modelled (Had) monthly climate data for the future time slice (2070–2099).

	Had original	Had annual	Had monthly	Had spatial	Had SD
Precipitation					
Mean	77.2	61.5	77.2	76.2	76.4
Median	42.8	6.2	7.8	30.0	25.0
Minimum	0.3	0.0	0.0	0.0	0.0
Maximum	335.7	436.5	547.3	592.9	730.4
SD	81.7	98.7	123.8	102.0	108.4
10th	0.4	0.5	1.1	0.4	1.9
90th	237.5	297.8	198.0	231.4	197.8
Temperature					
Mean	30.9	30.3	30.3	30.3	28.8
Median	30.2	29.6	29.9	29.8	28.0
Minimum	24.7	24.1	24.0	24.1	23.3
Maximum	37.7	37.1	36.6	36.7	36.6
SD	3.1	3.1	2.7	2.7	3.2
10th	27.4	26.8	27.0	27.1	25.0
90th	35.7	35.1	34.1	34.1	34.2

Observed Climate Change and Future Projections in the Seven River Basins

In order to get a general impression of historic climate variability and climate change, and of the future changes and variability as projected by climate model simulations, the past and coming 100 years were plotted in one graph for each river basin. Figure 2.8 depicts the observed precipitation and temperature record for the seven ADAPT basins over the period 1901–2000 and the projected climate change for the periods up to 2099 based on HadCM3 data and up to 2100 based on the ECHAM4 data, using the A2 and B2 SRES scenarios. Note that the observed data come from the CRU TS 2.0 database (Mitchell, T.D., Carter, T.R., Jones, P.D., Hulme, M. and New, M., 2003, unpublished), which is different from the CRU TS 1.2 data (New *et al.*, 2000) used in the analysis for the individual basins. This new dataset was not available at the start of the ADAPT project.

For constructing these figures, the GCM data were linearly stretched from the original size to 0.5° × 0.5°, after which the data were corrected using the statistical scaling method 2 mentioned earlier. The HadCM3 data were corrected relative to observed (CRU) record over the period 1961–1990, while the ECHAM4 data were corrected relative to the observed data over the period 1990–2000. Note that the axis for temperature has the same scale in each figure, whereas the axis for precipitation is different in each figure.

The HadCM3 data show that over the period 1950–1990 the climate model simulation for the A2 and B2 scenarios are equal, since the SRES scenarios assume the period before 1990 as the baseline, i.e. no change in greenhouse gas concentrations.

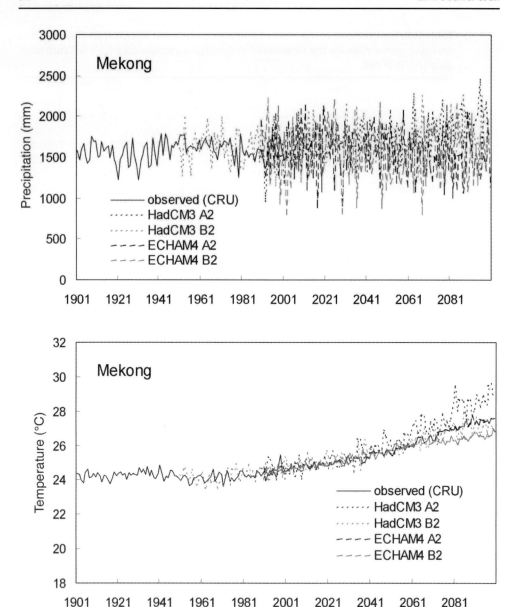

Fig. 2.8. Observed total annual precipitation and annual average temperature in the seven ADAPT river basins over the period 1901–2000 and projections for the periods up to 2099 based on the HadCM3 data and up to 2100 based on the ECHAM4 data. Observed data from CRU TS 2.0 (Mitchell, T.D., Carter, T.R., Jones, P.D., Hulme, M. and New, M., 2003, unpublished), projected data from IPCC-DDC (http://ipcc-ddc.cru.uea.ac.uk/).

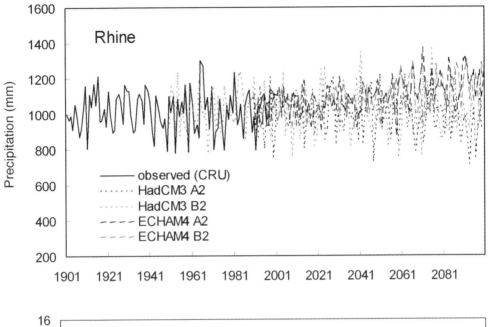

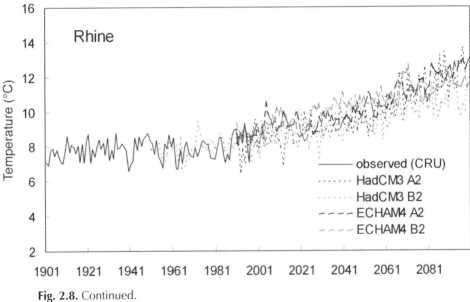

Fig. 2.8. Continued.

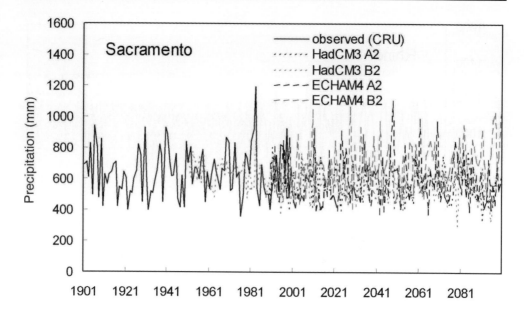

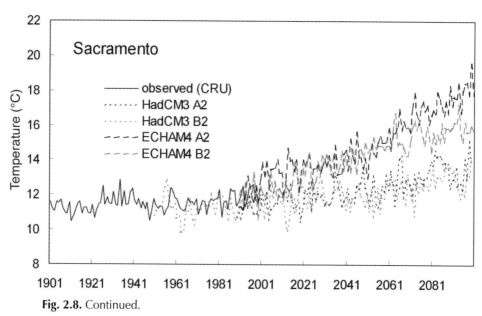

Fig. 2.8. Continued.

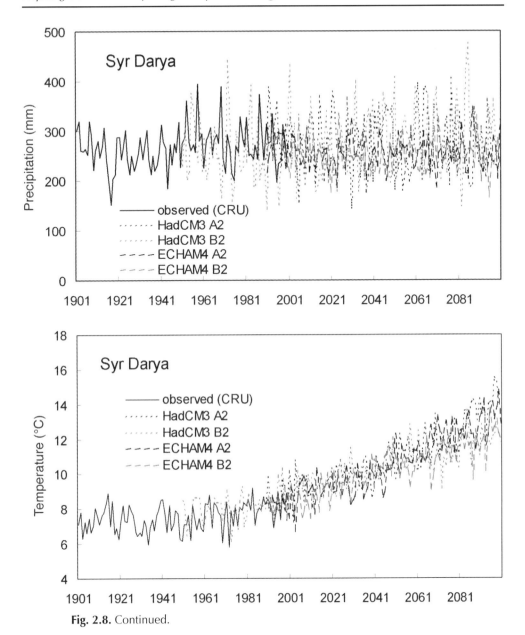

Fig. 2.8. Continued.

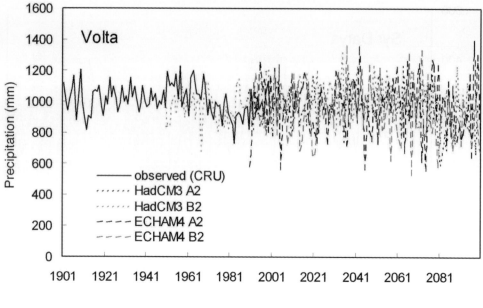

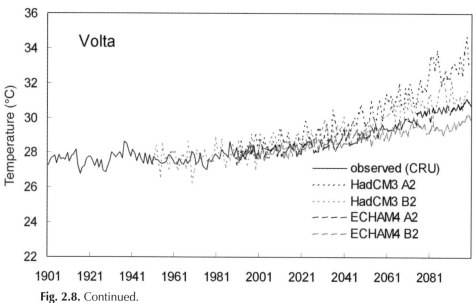

Fig. 2.8. Continued.

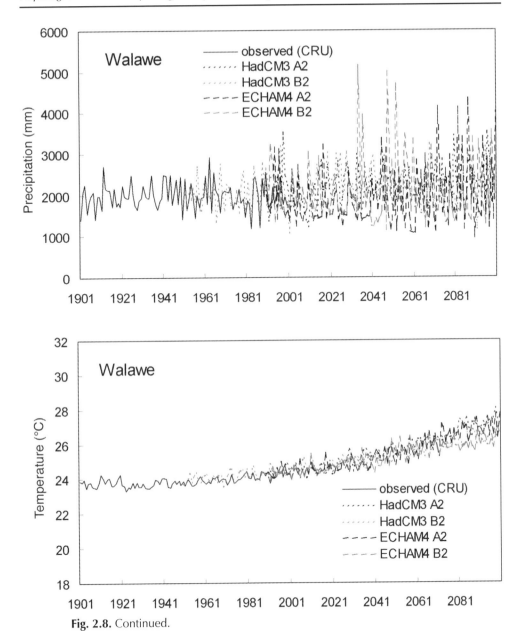

Fig. 2.8. Continued.

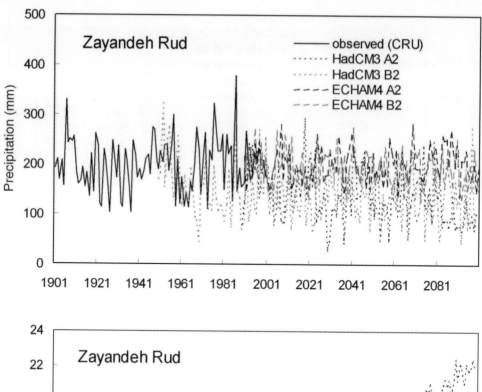

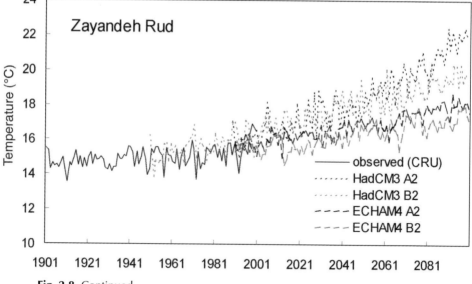

Fig. 2.8. Continued.

Table 2.3. Projected relative changes in precipitation (%) and absolute changes in temperature for the seven river basins by 2070–2099, relative to the period 1961–1990 (HadCM3 simulation) and relative to the period 1990–1999 (ECHAM4 simulation).

	Precipitation change [%]				Temperature change [°C]			
	HadCM3		ECHAM4		HadCM3		ECHAM4	
Basin	A2	B2	A2	B2	A2	B2	A2	B2
Mekong	17.2	5.6	−4.3	−0.2	3.9	2.8	2.6	1.9
Rhine	−7.0	−1.9	15.4	11.6	3.7	2.5	3.4	2.8
Sacramento	−9.7	−14.5	−11.3	10.9	2.2	1.6	5.4	3.7
Syr Darya	5.6	14.9	−4.5	−6.7	5.0	3.6	4.4	3.6
Volta	−7.1	0.4	2.2	−9.3	4.7	3.2	2.5	1.5
Walawe	17.2	7.3	7.2	−1.6	2.7	2.0	2.4	1.7
Zayandeh	−12.2	−5.5	0.4	−6.6	5.1	3.6	2.0	1.4

The graphs not only show the observed and projected changes, but also to what extent the models agree on the change in precipitation and temperature, based on the same set of emission scenarios.

Table 2.3 lists the projected average changes in precipitation and temperature for the period 2070–2099 in the seven river basins. From this table, the differences between the two scenarios become clear. In general, the B2 scenario holds a somewhat milder future than the A2 scenario, under which higher temperatures and more radical changes in precipitation occur. It also becomes clear that the two models do only seldom agree on the direction and amount of change in precipitation. This is for a major part due to the fact that the basins are spatially relatively small features, and the GCMs are apparently not capable of simulating the same changes on a small scale. Moreover, the geographically complex location of some of the basins with respect to mountainous areas and large oceans (e.g. Sacramento, Volta and Mekong River Basins) makes consistent projections more difficult.

Table 2.4 depicts the direction and magnitude of change in precipitation and temperature by 2070–2099, as well as the consistency between the two climate models. For the Sacramento and Zayandeh Basins, a decrease in average annual precipitation is projected under a single scenario for both models. For the Walawe Basin under the A2 scenario, the average annual precipitation amount is projected to increase. Future temperature changes are much more consistently simulated for most of the basins. In particular, large increases are projected for the Syr Darya and the Rhine Basins.

Conclusions

In this chapter, four statistical downscaling techniques were introduced and one spatial downscaling technique (Kriging). Statistical methods 1 and 2 are simple methods that correct future projections based on annual and monthly averages respectively. Method 3 adds to method 2 a correction across all spatial raster cells.

Table 2.4. Projected relative changes in precipitation and absolute temperature change and consistency between the two models (HadCM3 and ECHAM4) for two scenarios for the seven river basins. The signs + + and − denote a change in precipitation larger than 10%, +/− a change between 5 and 10%. 'i' denotes inconsistency, which is defined as a difference in precipitation change larger than 10%, or a difference in temperature change of more than 1.5°C between the two model projections.

	Precipitation		Temperature	
Basin	A2	B2	A2	B2
Mekong	i	i	2.6–3.9	1.9–2.8
Rhine	i	i	3.4–3.7	2.5–2.8
Sacramento	−/−−	i	i	i
Syr Darya	i	i	4.4–5.0	3.6
Volta	i	i	i	i
Walawe	+/++	i	2.4–2.7	1.7–2.0
Zayandeh	i	−	i	i

Method 4 is a method that corrects the GCM outputs using the standard deviation of the measured data.

All downscaling techniques were applied to the baseline period (1961–1990). The results were compared in graphs and tables using statistical test parameters. Looking at the RMS errors, methods 1 and 2 perform worst and methods 3 and 4 perform best. All four methods are able to correct the mean values of the GCM projections for temperature and precipitation to the measured data. However, after correction, the corrected extreme values from methods 1 and 2 in particular do not match the measured values well.

It is therefore recommended to use either method 3 or method 4 for downscaling. It is expected that errors propagate and increase when using the corrected sets for additional modelling (such as hydrological modelling). Hence, using the best method with the best fit will lower final errors in model simulations using the downscaled data.

References

Alcamo, J., Döll, P., Kaspar, F. and Siebert, S. (1997) *Global Change and Global Scenarios of Water Use and Availability: An Application of WaterGAP 1.0.* Report A9701, Centre for Environmental Systems Research, University of Kassel, Kassel, Germany.

Allen, M., Raper, S. and Mitchell, J. (2001) Uncertainty in the IPCC's Third Assessment Report. *Science* 293, 430–433.

Arnell, N.W., Bates, B., Lang, H., Magnuson, J.J., Mulholland, P., Fisher, S., Liu, C., McKnight, D., Starosolszky, O., Taylor, M., Aquize, E., Arnott, S., Brakke, D., Braun, L., Chalise, S., Chen, C., Folt, C.L., Gafny, S., Hanaki, K., Hecky, R., Leavesly, G.H., Lins, H., Nemec, J., Ramasastri, K.S., Somlyódy, L. and Stakhiv, E. (1996) Hydrology and freshwater ecology. In: Watson, R.T., Zinyowera, M.C. and Moss, R.H. (eds) *Climate Change 1995 – Impacts, Adaptations and Mitigation of Climate Change: Scientific-Technical Analyses. Contribution of Working Group II to the Second Assessment of the*

Intergovernmental Panel on Climate Change. Cambridge University Press, Cambridge, pp. 325–363.

Arnell, N.W., Liu, C., Compagnucci, R., Da Cunha, L., Howe, C., Hanaki, K., Mailu, G., Shiklomanov, I.A., Döll, P., Becker, A. and Zhang, J. (2001) Hydrology and water resources. In: McCarthy, J.J., Canziani, O.F., Leary, N.A., Dokken, D.J. and White, K.S. (eds) *Climate Change 2001 – Impacts, Adaptation and Vulnerability. Contribution of Working Group II to the Third Assessment Report of the Intergovernmental Panel on Climate Change.* Cambridge University Press, Cambridge, pp. 191–233.

Burough, P.A. (1986) *Principles of Geographical Information Systems for Land Resources Assessment.* Clarendon Press, Oxford.

IPCC-TGCIA (1999) *Guidelines on the Use of Scenario Data for Climate Impact and Adaptation Assessment, Version 1.* Carter, T.R., Hulme, M. and Lal, M. (eds) Intergovernmental Panel on Climate Change, Task Group on Scenarios for Climate Impact Assessment.

IPCC (2000) *Emission Scenarios: A Special Report of Working Group III of the Intergovernmental Panel on Climate Change.* Nakićenović, N. and Swart, R. (eds) Cambridge University Press, Cambridge.

IPCC (2001) *Climate Change 2001 – The Scientific Basis. Contribution of Working Group I to the Third Assessment Report of the Intergovernmental Panel on Climate Change.* Houghton, J.T., Ding, Y., Griggs, D.J., Noguer, M., van der Linden, P.J., Dai, X., Maskell, K. and Jonhnson, C.A. (eds) Cambridge University Press, Cambridge.

Jones, P.D. (1994) Hemispheric surface air temperature variations: a reanalysis and update to 1993. *Journal of Climate* 7, 1794–1802.

McGuffie, K. and Henderson-Sellers, A. (1997) *A Climate Modelling Primer,* 2nd edn. John Wiley & Sons, Chichester.

New, M., Hulme, M. and Jones, P. (2000) Representing twentieth-century space-time climate variability. Part II: Development of 1901–96 monthly grids of terrestrial surface climate. *Journal of Climate* 13, 2217–2238.

Parry, M.L. (2002) Scenarios for climate impact and adaptation assessment. *Global Environmental Change* 12, 149–153.

Wigley, T.M.L. and Raper, S.C.B. (2001) Interpretation of high projections for global-mean warming. *Science* 293, 451–454.

Wilby, R.L. and Wigley, T.M.L. (1997) Downscaling general circulation model output: a review of methods and limitations. *Progress in Physical Geography* 25, 530–548.

3 Adaptation Strategies to Climate Change to Sustain Food Security

PETER DROOGERS,[1] JOS VAN DAM,[2] JIPPE HOOGEVEEN[3] AND
RONALD LOEVE[1]

[1]FutureWater, Arnhem, The Netherlands; [2]Wageningen University,
Wageningen, The Netherlands; [3]FAO, Rome, Italy

Introduction

Producing sufficient food to sustain the growing world population is one of the key challenges for now and for the coming decades. According to a study recently presented by the Food and Agriculture Organization of the United Nations (FAO), about 799 million people in the developing world do not have enough to eat and another 41 million in industrialized countries and countries in transition also suffer from chronic food insecurity (FAO, 2002a). More than half of these undernourished people (60%) are found in Asia, while sub-Saharan Africa accounts for almost a quarter (23%). In terms of the percentage of undernourished people of the total population, the highest incidence is found in sub-Saharan Africa, where it was estimated that one-third of the population (34%) was undernourished in 1997–1999. Sub-Saharan Africa is followed by South Asia, where 24% of the population is undernourished (Fig. 3.1). However, significant progress has been made over the last two decades: the incidence of undernourishment in developing countries has decreased from 29% in 1979–1981 to 17% in 1997–1999.

The World Bank estimates of the prevalence of extreme poverty are based on the distribution of household expenditure on consumables (Chen and Ravallion, 2000), while the FAO estimates are based on the distribution of household food consumption and availability. World Bank indices indicate in general higher undernourishment than FAO ones (Fig. 3.1), but there is a positive and close relationship between food consumption and expenditure on consumables in low-income households.

Vörösmarty *et al.* (2000) argue that the impact of climate change on water and food will be relatively small in comparison to the impact of changes in population and socio-economic projections. However, changes in population and socio-economic issues are gradual and therefore easier to cope with. Climate change includes such a gradual change as well, but is expected to increase extremes substantially.

This chapter will concentrate on the impact of climate change on food and water issues. An overview of global issues and trends will be followed by a more in-depth

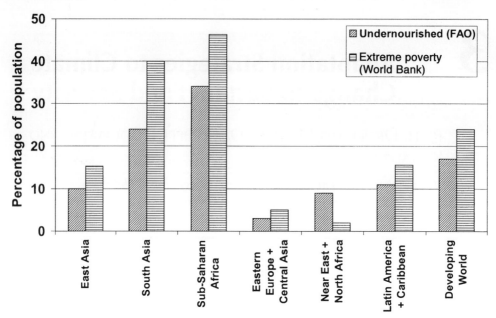

Fig. 3.1. Incidence of undernourishment and poverty according to FAO and World Bank (Chen and Ravallion, 2000) definitions.

analysis of field-scale impact and adaptation strategies for the seven basins in the ADAPT context. The field scale is of paramount importance as this is the scale at which food is actually produced. The relation with the basin scale will be discussed, but the subsequent basin chapters (Chapters 5–11) will provide more detail. The field-scale analysis will be oriented towards comparison of the seven basins.

Global Food Issues

Current situation

Food trends in the developing and developed world have shown very different patterns over the last 30 years. Overall, food production is steadily increasing, especially in the developing world (Fig. 3.2, top). Food production per capita, however, is not increasing in the same manner, especially in the developing world (Fig. 3.2, bottom).

An enormous amount of water is required to produce food. FAO's AQUASTAT indicates that from all water diverted, 69% is used in irrigation and 10% and 21% for domestic use and industry, respectively (FAO, 2003b). These data are based on withdrawals and not on consumption. Since return flows from domestic use and industry are in general high and can be reused, actual consumption from agriculture will be even higher in comparison to other sectors. It is estimated that in developing countries irrigated agriculture accounts for about 20% of all cultivated land. Of the total amount of agricultural production, 40% originates from irrigated land, while for

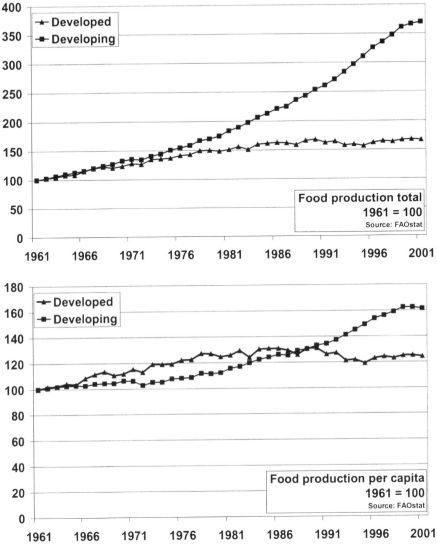

Fig. 3.2. Global trends in food production over the last 40 years expressed as total food produced (top) and per capita (bottom). Depicted figures are indexed with the period 1961 set as 100.

cereals as much as 60% of production originates from irrigated agriculture (FAO, 2002b).

A study recently presented by Shiklomanov (2003) used a modern assessment framework to assess the state of the world's water resources. The study focuses not only on water availability now and in the future, but also includes estimates of water withdrawals and consumption over the last 100 years. Figures presented in this study for the year 2000 compare quite well with the AQUASTAT data that show a withdrawal for irrigation of about 66% of total diversions (Fig. 3.3). In terms of consumption, the

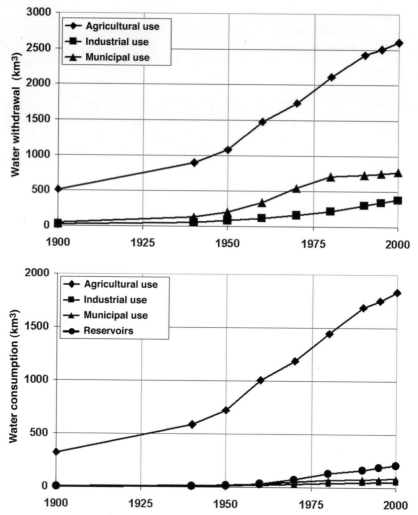

Fig. 3.3. Dynamics of water withdrawals (top) and consumption (bottom) per sector. (Source: Shiklomanov, 2003.)

agricultural sector is responsible for 84%, while figures for domestic use, industry and reservoirs are 4%, 2% and 10%, respectively. The substantial consumption by reservoirs is often ignored, but is estimated to be even higher than consumption by domestic use and industry together. The total storage capacity of all reservoirs in the world is approximately 6000 km³, almost three times the total annual withdrawals in the world.

Over the last 100 years agriculture has always been the dominant user of diverted water (Fig. 3.3). Since 1950, diversions for domestic use and industry are rising, but consumption is still low in comparison to agriculture. However, it should be considered that water quality requirements will be different for the three sectors considered. For example, water quality requirements will be much stricter for domestic use than

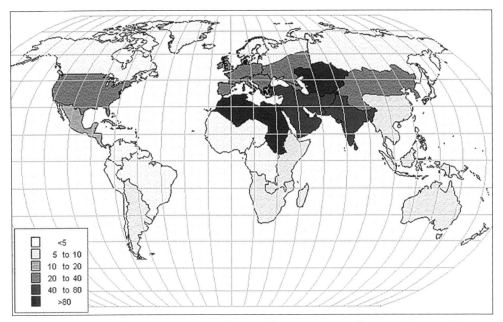

Fig. 3.4. Percentage of water consumption from total water resources by natural-economic regions of the world in 2000. (Source: Shiklomanov, 2003.)

for agriculture. Looking at the spatial distribution, the drier regions of the Northern Hemisphere currently consume a vast amount of their fresh water resources (Fig. 3.4). It is alarming, specifically in China, India and Pakistan, that population growth is high and it is uncertain whether food security can be sustained and enhanced considering future water requirements.

Projections for the future

Seckler *et al.* (1999) estimates that by 2025 cereal production will have to increase by 38% to meet world food demands. The World Water Vision, an outcome of the Second World Water Forum in The Hague in 2000, estimated a similar increase of 40% based on various projections and modelling exercises. These figures are based on an econometric model showing that grain production will increase about 2% per year for the 2000–2020 period (Koyama, 1998). Following UN mid-range population estimates of 8.9 billion people, combined with the minimum caloric requirement of 2200 calories per day, means that a total of about 20 trillion consumable calories have to be produced. Current levels are at about 14 trillion calories, which means an increase of 42% is needed. However, given the range in population estimates provided by the UN, this figure can be between 14% and 71%.

The increase in food production and related water requirements coincides with an alarming increase in water scarcity. One-third of the world's total population of 5.7 billion lives under conditions of relative water scarcity and 450 million people are living under severe water stress (UNEP, 1998). This relative water scarcity and severe water stress are defined using the Relative Water Demand (RWD), expressed as the fraction of water demand over water supply. A RWD larger than 0.2 indicates relative water scarcity, while a RWD greater than 0.4 indicates severe water stress. However, the UN values are based on national-level totals, ignoring the fact that especially in bigger countries, huge spatial differences can occur. Vörösmarty *et al.* (2000) show that, including these in-country differences, 1.8 billion people live in areas with severe water stress. Using a global water model and projections for climate change, population growth and economic growth, they concluded that the number of people living in severe water stress will have grown to 2.2 billion by the year 2025.

Food Production and Climate Change

Global perspectives

The previous section showed that the question whether sufficient food can be produced is associated with many uncertainties. On top of this will be the impact of climate change. It has been demonstrated that the impact of climate change will be minor in comparison to changes in population and socio-economic issues (e.g. Vörösmarty *et al.*, 2000). One important issue ignored in these analyses is that changes in population and socio-economic issues are gradual. Climate change also includes such gradual change, but more important is that extreme weather events are expected to increase substantially. These extremes are very difficult to cope with in the context of food production in particular.

A recent FAO study (FAO, 2002b) claims that in the next three decades, climate change is not expected to depress global food availability, but it may increase the dependence of developing countries on food imports and accentuate food insecurity for vulnerable groups and countries. The study indicates that the percentage of undernourished people will fall from the current figure of 17% to 11% in 2015 and 6% by 2030. This seems a positive projection. However, the target set by the World Food Summit in 1996 and reiterated as a Millennium Goal – of halving the number of undernourished people by 2015 – is not going to be met unless major corrective efforts are made. In terms of water requirements, the projections for developing countries indicate a 14% increase in water withdrawals for irrigation by 2030.

In contrast to these figures, results presented by the International Food Policy Research Institute (Rosegrant *et al.*, 2002) show that the increase in water withdrawals for irrigation can only be 4%, mainly as a result of water shortages. In their business as usual strategy, farmers will produce 10% less cereals, pushing up food prices sharply. The authors continue that it is possible to envision a sustainable water strategy that would dramatically increase water allocated to environmental uses, connect all urban households to piped water, and achieve higher per capita domestic water consumption, while maintaining food production at the levels described in the business as usual strategy.

The International Institute for Applied Systems Analysis (IIASA), together with FAO, presented a study based on their GAEZ approach (Global Agro-Ecological Zoning), which has a strong focus on lands, including several Global Circulation

Model (GCM) projections (Fischer *et al.*, 2001). The evaluation of the impact of climate change on production, consumption and trade of agricultural commodities, in particular on staple food, was carried out with a large number of experiments that relate to four aspects: magnitude of climate change for different future socio-economic and technical development paths; uncertainty of results in view of differences in climate projections of different GCM groups; robustness of results with regard to altered economic growth assumptions; and sensitivity of results to different assumptions with regard to physiological effects of atmospheric CO_2 enrichment on yields. Some 50 simulation experiments were carried out, including three separate snapshots of climatic change for the 2020s, 2050s and 2080s. Their results indicate that the impacts of climate change on crop production are geographically unevenly distributed. Developed countries experience an increase in productivity. In contrast, developing regions suffer a loss in cereal productivity in all estimates. Within the group of developed countries, gains of 3–10% in cereal productivity occur for North America, and similar for the Former Soviet Union. Western Europe suffers losses in most projections of up to 6%.

The IIASA study continues that in terms of food security fairly robust conclusions emerge from the analysis of climate-change impacts. First, climate change will most likely increase the number of people at risk of hunger. Secondly, the importance and significance of the climate-change impact on the level of undernourishment depends largely on the level of economic development assumed in the strategies.

Impact of CO_2 on crop growth

Crop production is affected by the atmospheric carbon dioxide (CO_2) level. Photosynthetically active radiation (PAR) is used by the plant as energy in the photosynthesis process to convert CO_2 into biomass. It is important to make a distinction in this process between C3 and C4 plants. The difference between C3 and C4 plants is the way the carbon fixation takes place. C4 plants are more efficient and the loss of carbon during the photorespiration process is negligible for C4 plants. C3 plant may lose up to 50% of their recently fixed carbon through photorespiration. This difference has suggested that C4 plants will not respond positively to rising levels of atmospheric CO_2. However, it has been shown that atmospheric CO_2 enrichment can, and does, elicit substantial photosynthetic enhancements in C4 species (Wand *et al.*, 1999). Examples of C3 plants are potatoes, sugarbeet, wheat, barley, rice and most trees except mangrove. C4 plants are mainly found in the tropical regions and some examples are millet, maize and sugarcane. A third category is the so-called CAM (crassulacean acid metabolism) plants, which have an optional C3 or C4 pathway of photosynthesis, depending on conditions: examples are cassava, pineapple and onions.

Modelling studies based on detailed descriptions of crop growth processes also indicate that biomass production and yield will increase under elevated CO_2 levels. For example, Rötter and van Diepen (1994) showed that potential crop yields for several C3 plants in the Rhine Basin will increase by 15–30% in the next 50 years as a result of increased CO_2 levels. According to their model, the expected increase in yield for maize, a C4 plant, will be only 3%, indicating that their model was indeed based on the assumption that C4 species do not benefit from higher CO_2 levels.

Table 3.1. Increase of potential crop growth as a result of enhanced CO_2 levels in percentages. A2 and B2 are the IPCC climate scenarios. (Source: CSCDGC, 2002.)

Crop	Period	A2 (%)	B2 (%)
Rice	2010–2030	20	10
	2070–2100	40	20
Cotton	2010–2030	25	13
	2070–2100	50	25
Wheat	2010–2030	20	10
	2070–2100	40	20
Maize	2010–2030	10	5
	2070–2100	20	10
Beet	2010–2030	10	5
	2070–2100	20	10
Tomato	2010–2030	15	8
	2070–2100	30	15

In addition to these theoretical approaches, experimental data have been collected to assess the impact of a CO_2-enriched atmosphere on crop growth. A vast number of experiments have been carried out over the last decades to quantify the impact of increased CO_2 levels on crop growth. The Center for the Study of Carbon Dioxide and Global Change in Tempe, Arizona (CSCDGC, 2002) has collected and combined results from this kind of experiments (Table 3.1). The example of rice shows that average biomass increases are 31% for increases in atmospheric CO_2 concentrations of 300 ppm, but the variation is substantial (Fig. 3.5).

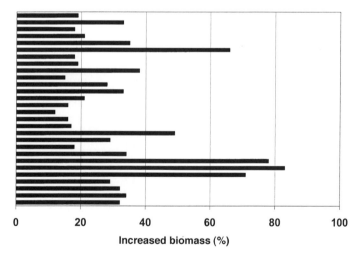

Fig. 3.5. Example of the impact of elevated CO_2 levels on crop production according to experiments. Data are from 26 greenhouse experiments with CO_2 concentrations of 650 ppm for rice. (Source: CSCDGC, 2002.)

Impact of temperature on crop growth

In general, higher temperatures are associated with higher radiation and higher water use. Two effects of temperature are distinguished: (i) the physiological effects (at the level of plants and plant organs); and (ii) the crop-ecosystem effects (at the level of the field or at the region). These effects have both positive and negative impacts, largely depending on local conditions and plant species (Bazzaz and Sombroek, 1996).

At the plant level, it is expected that rising temperatures would diminish the yield of some crops, especially if night temperatures are higher (Kukla and Karl, 1993). However, as long as plants do not get overheated, some positive effects on crop growth can also be expected. A negative aspect is that higher cold-season temperatures may lead to earlier ripening of annual crops, diminishing yield per crop. An adaptation option to cope with this effect is to allow locally for the growth of more crops per year due to the lengthening of the growing season.

At the crop-ecosystem level, it is expected that crop yields will be enhanced in the northern regions of the former Soviet Union, Canada and Europe. The cereal-growing belts of North America might shift northwards by several hundred kilometres for every degree Celsius rise in temperature. The predicted yield increases in the higher latitude regions are primarily due to a lengthening of the growing season and the mitigation of negative cold weather effects on plant growth (Parry and Rosenzweig, 1993). Negative effects on crop and livestock productivity are expected in northern middle latitude countries like the USA, Western Europe and most of Canada's currently productive agricultural regions. This is due to a shortening of the growing period caused by increased temperatures and evapotranspiration rates (Tobey *et al.*, 1992; see also Chapter 4).

Adaptation and Food Security

Smit and Skinner (2002) showed that many different adaptation measures exist in the agricultural sector. Adaptations range from farmers changing management practices, timing of operations or crop choice, to public agencies investing in technological developments or irrigation schemes, or modifying support programmes and information sharing or early warning systems. However, little information exists on the conditions under which adaptation measures are likely to be adopted. The limited research to date indicates that producers rarely respond to climate change alone, and that adaptation to climate change risks would be undertaken as part of ongoing production and risk management decision-making.

In an effort to better understand the potential global impacts of climate change on agriculture, a major study was performed concerning the effects of changing temperature and precipitation regimes and increased CO_2 concentrations on crop production and its economic implications (Rosenzweig *et al.*, 1993). The central aim of the study was to provide an assessment of potential climate change impacts on world crop production, including quantitative estimates of changes of: major food, cash and industrial crop yields; prices; trade and risk of hunger. Three climate periods and three levels of farmer adaptation options to climate change were assumed.

- No adaptation.
- Level 1 adaptation: changes that imply small additional costs to farmers and no necessary policy changes, such as shifts in planting dates, variety and crop, and increases in water application to irrigated crops.
- Level 2 adaptation: higher order adaptations that imply significant additional costs to farmers, such as large shifts in crop production timing, increased fertilizer application, installation of irrigation systems, and development of new varieties, and/or changes in policy.

Globally, both minor and major levels of adaptation help restore world production levels, compared to the climate change scenarios with no adaptation. Average global cereal production decreases by up to about 5% from the reference case under Level 1 adaptations. These involve shifts in farm activities that are not very disruptive to regional agricultural systems. With adaptations implying major changes, global cereal production responses range from a slight increase to a slight decrease ($+1\%$ to -2.5%).

A World Bank study (Mendelsohn and Dinar, 1999) stated that because most developing countries depend heavily on agriculture, the effects of global warming on productive croplands are likely to threaten both the welfare of the population and the economic development of the countries. Tropical regions in the developing world are particularly vulnerable to potential damage from environmental changes because the poor soils that cover large areas of these regions mean much of the land is already unusable for agriculture. Although agronomic simulation models predict that higher temperatures will reduce grain yields as the cool wheat-growing areas get warmer, they have not examined the possibility that farmers will adapt by making production decisions that are in their own best interests. A set of models examines cross-sectional evidence from India and Brazil and finds that even though the agricultural sector is sensitive to climate, individual farmers do take local climates into account, and their ability to do so will help to mitigate the impacts of global warming.

A study focusing on the USA emphasized that farmers have many adaptation options, such as changing planting and harvest dates, rotating crops, selecting crops and crop varieties for cultivation, irrigation, using fertilizers and choosing different tillage practices (Adams *et al.*, 1999). These adaptation strategies can reduce potential yield losses from climate change and improve yields in regions where climate change has beneficial effects. At the market level, price and other changes can signal further opportunities to adapt as farmers make decisions about land use and which crops to grow. Thus, patterns of food production respond not only to biophysical changes in crop and livestock productivity brought about by climate change or technological change, but also to changes in agricultural management practices, crop and livestock prices, the cost and availability of inputs, and government policies. In the longer term, adaptations include the development and use of new crop varieties that offer advantages under changed climates, or investments in new irrigation infrastructure as insurance against potentially less reliable rainfall. The extent to which opportunities for adaptation are realized depends upon a variety of factors such as information flow, access to capital and the flexibility of government programmes and policies.

A major obstacle for farmers is the expected increase in successive years of overly dry and/or wet spells. A poor farmer might overcome a 1-year drought followed by a normal year, but a period of 2 or more years of drought, even followed by a longer period of normal years, will be catastrophic to this farmer. Crop insurance programmes for small-scale, traditional or subsistence farmers can be developed and implemented in a viable and sustainable manner. In addition, micro-insurance provides one of the few alternatives to the rural populations and the poor as a means of social protection (CERUDEB, 2002).

As mentioned before, there exist many lists of possible adaptation measures, initiatives or strategies that have a potential to moderate impacts, if they were implemented. Such possible adaptations are based on experience, observation and speculation about alternatives that might be created and they cover a wide range of types and can take numerous forms (UNEP, 1998). It is important to make a distinction between the different players as to who can do what. In addition to these different players, the location within a basin is also equally important for defining the most adequate adaptation strategy.

Field-scale Adaptation in Seven Contrasting Basins

Although the ADAPT project focused mainly on the basin scale, attention has been paid to assessing the impacts of climate change on food at the local to field scale, including developing local response strategies. For this, the SWAP (Soil–Water– Atmosphere–Plant; van Dam *et al.*, 1997) model was applied for calculating yields and production for two representative crops in each of the seven basins. This was done for three periods: 1961–1990, 2010–2039 and 2070–2099. Moreover, for the periods 2010–2039 and 2070–2099, the impacts of climate change on yield and production were explored under:

1. The current climate (1961–1990).
2. Climate change for 2010–2039 and 2070–2099 using the A2 and B2 scenarios from the HadCM3 climate model (see Chapter 2) and business as usual strategy.
3. Under climate change scenarios for 2010–2039 and 2070–2099 and two adaptation strategies (see the previous section).

The FAO/UNESCO Digital Soil Map of the World was used to derive soil physical parameters, required for simulating soil water processes in the unsaturated–saturated zone. Although locally more detailed soil maps might exist, we chose to use this global dataset to ensure that simulation results will not be a function of the different approaches used to generate datasets. Texture data, organic matter content and bulk density were used to derive soil physical functions (retention curve and hydraulic conductivity) by applying pedo-transfer functions as developed by Wösten *et al.* (1998).

The agro-hydrological analysis at field scale is performed using the SWAP 2.0 model (van Dam *et al.*, 1997). SWAP is a one-dimensional physically based model for water, heat and solute transport in the saturated and unsaturated zones, and also includes modules for simulating irrigation practices and crop growth. The water transport module in SWAP is based on the well-known Richards' equation, which is a combination of Darcy's law and the continuity equation. A finite difference solution scheme is used to solve Richards' equation. Crop yield can be computed using a simple crop growth algorithm based on Doorenbos and Kassam (1979) or by using a detailed crop growth simulation module that partitions the carbohydrates produced between the different parts of the plant, as a function of the different phenological stages of the plant (van Diepen *et al.*, 1989). Potential evapotranspiration is partitioned into potential soil evaporation and crop transpiration using the leaf area index. Actual transpiration and evaporation are obtained as a function of the available soil water in the top layer or the root zone for, respectively, evaporation and transpiration. Finally, irrigation can be prescribed at fixed times, can be scheduled according to different criteria, or a combination of both can be used. A detailed description of the model can be found in van Dam *et al.* (1997).

Basin-specific data

Mekong

For the Mekong Basin, the two major crops have been selected for the analysis: rice and maize. Thailand and Vietnam, two riparian countries in the basin, are rice exporters numbers one and two, respectively, in the world market. The Mekong River Delta has three major cropping seasons for rice: winter–spring or early season with full irrigation at the end of crop season (November to February), summer–autumn or mid-season with supplementary irrigation at the beginning of crop season (June to September), and main rain-fed season (June to December), the long-duration wet season crop. The largest rice area is cropped during the summer–autumn season. The rice yield is highest in the winter–spring season, and lowest in the main rainy season. Farmers in this region adopt a direct-seeding method of crop establishment to save labour costs. In the Mekong River Delta of South Vietnam there are about 70,000–100,000 ha of land which flood every year. After the flood season, early maturing corn can be planted on untilled wetlands in January, and harvested in April. This practice was first developed in some provinces in Vietnam, and has been extended to thousands of hectares. Maize gives average yields of 3.0–3.5 t/ha, while skilled farmers attain yields of 5 t/ha (FFTC, 2003).

Rhine

Sugarbeet and wheat are grown extensively in the Rhine Basin. Sugarbeet is planted in April and harvested in October. Potential yields are around 60 t/ha and average crop prices are about €40 per tonne, depending on the sugar content of the beet (IRS, 2003). These high yields combined with a reasonable price makes sugarbeet an attractive crop. Somewhat less profitable, but essential in the crop rotational pattern, is wheat. In most cases varieties requiring a cold period are used and are therefore planted around October or November. World prices for wheat have been reasonably high recently, at around €0.13 per kilogram, but farmers in the Rhine Basin can always rely on an agreed minimum price from the European Union, which is currently at €0.10 per kilogram (EC, 2003; FAO, 2003a).

Sacramento

Rice production in the Sacramento Basin is of major interest from an agricultural as well as environmental point of view. Rice is associated with heavy use of the scarce water resources in the basin, but also generates substantial economic benefits. From the environmental point of view, these paddy fields are valuable wetlands, but at the same time are associated with water quality degradation. The second crop considered in the Sacramento Basin is tomato. The area of tomatoes is steadily increasing and the product is used for fresh market as well as processed tomatoes. A major constraint in tomato production, besides land and water resources, is sufficient labour. Tomato prices can fluctuate substantially and, like other commodities, depend highly on the total production (Hartz, 2003).

Syr Darya

The Syr Darya region is traditionally a major cotton area, and could be considered as the cotton belt of the former Soviet Union. The collapse of the Soviet Union has induced a large change in cropping patterns, where cotton is partly replaced by wheat and other cereals (Lindeman, 2003). Cotton is associated with high water consumption and with the well-known dramatic water-level decline of the Aral Sea. The second crop selected for the basin is wheat.

Volta

The two selected crops for the Volta Basin are rice and maize. Although rice consumption is at a relatively low level, with, for example, less than 10 kg milled rice per person per year in Ghana, it is considered to be an important crop. Average yield levels are low, as a result of many factors such as inadequate expertise, low investment options, poor maintenance of irrigation and unsuitable rice varieties. Maize was selected as the other crop included in the analysis.

Walawe

Rice is the major food crop in the Walawe Basin, from which a considerable amount is exported to Colombo, the commercial capital of Sri Lanka. Normally two crops can be grown during the year, one in the Maha (November to March) and one in the Yala (May to September) season. Rice is grown in the low lands with an area of about 30,000 ha. Rice is such a dominant crop in the basin that all other crops are normally referred to as OFC, other field crops. These are grown on small plots, the so-called gardens. As a second crop to represent these OFC in this study, vegetables were selected.

Selection of adaptation measures

The adaptation strategies evaluated here originate from discussions with stakeholders from the seven basins about the most likely and interesting options to explore (see earlier). Farmers' main options to adapt are: conversion from irrigation to rain-fed or vice versa, installation of drainage, a change in cropping patterns, salinity control and intensification in general. The latter is of paramount importance for developing countries, since the gap between obtainable and existing yields is still enormous (Fig. 3.6). Water managers at the basin level can take decisions that can minimize the adverse impact of climate change or, in cases where the impact is positive, make better use of these positive impacts. While farmers' adaptation is limited to only the agricultural sector, water managers' decisions are related to water allocation between and within

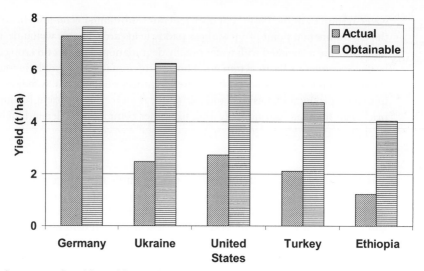

Fig. 3.6. Exploitable yield gaps for wheat: actual versus obtainable yields for some selected countries. (Source: Fisher *et al.*, 2002.)

different sectors. Adaptation strategies as studied at the field scale should also be considered in a basin-wide context. Reduced irrigation, as explored for a couple of crop–basin combinations, is not an adaptation strategy at the field scale to overcome the negative impact of climate change, but can be upscaled to a basin-wide strategy.

The number of crops to evaluate was limited to two and the adaptation strategies were limited to three. This was essential to minimize the number of years to evaluate, which is with the current defined options already over 15,000 years.

Results

Overall, the general picture is that crop yields will be higher in the future (Fig. 3.7), but that variation in yields between years (Fig. 3.8) will increase as well. This indicates that concerns are predominantly related to variations in food security expressed by farmers' income, and not to total food production. It should be noted that most of this increase in total production is an effect of the increase in CO_2, as presented in Table 3.1. There is still considerable debate on the validity of these data, as discussed earlier.

In terms of water resources, the total amount of water consumed (Fig. 3.9) is essential. A distinction has been made between total consumed water and so-called productively consumed water. The latter relates only to crop transpiration, while the former includes soil evaporation as well. This distinction is important since transpiration is associated with a beneficial consumption of water (crop per drop), while the latter can be seen as a real loss of water. However, a certain amount of soil evaporation it is not completely unavoidable. Important to the discussion on this productivity is that this is only relevant in cases where water is scarce. In this study percolation is not considered as water consumption, since most of this water is groundwater recharge and can be reused.

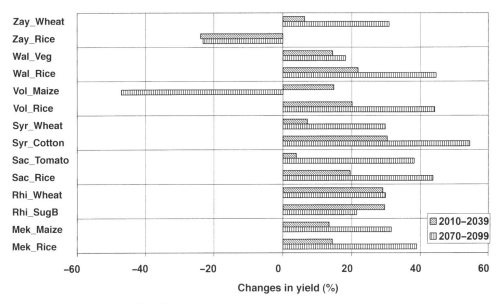

Fig. 3.7. Changes in yield for the periods 2010–2039 and 2070–2099 as compared to the baseline 1961–1990. Values for HadCM3 A2 climate change projections and business as usual strategy.

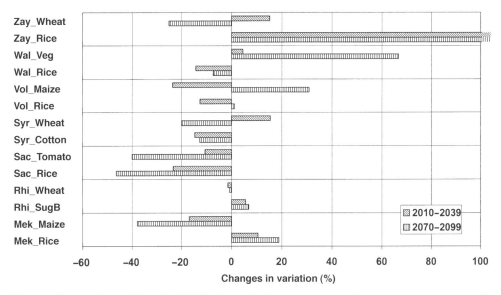

Fig. 3.8. Changes in variation in yield for the periods 2010–2039 and 2070–2099 as compared to the baseline 1961–1990. Values for HadCM3 A2 climate change projections and business as usual strategy.

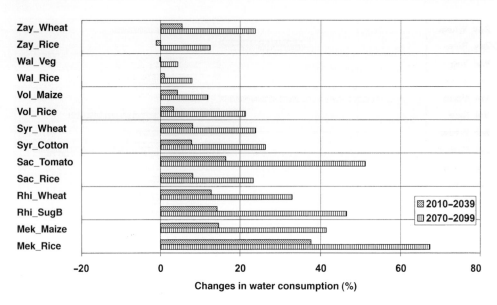

Fig. 3.9. Changes in water consumption for the periods 2010–2039 and 2070–2099 as compared to the baseline 1961–1990. Values for HadCM3 A2 climate change projections and business as usual strategy.

Water Productivity (WP), expressed as dollars gross return per m³ of water consumed (total consumed, so actual evapotranspiration), for each crop–basin combination was also calculated (Fig. 3.10). This is affected by local crop prices, but since these data were hard to obtain we have elected to use average world market prices over the last 5 years. Deviations are possible, since for some crops special varieties are grown that have higher local market values than world market ones. Overall, WP values range from a low $0.01/m³ to almost $1/m³.

A detailed description of results for the seven basins can be found elsewhere (Droogers and van Dam, 2004) and we will focus here on the inter-comparison of the seven basins, with some specific conclusions per basin.

Regarding the field scale inter-comparisons, one of the most striking conclusions that the overall picture of the impact of climate change on crop yields is positive (Fig. 3.7). In the business as usual option, expected yields are higher for all basins except one basin–crop combination for the period 2010–2039 and two basin–crop combinations for the period 2070–2099, as can be seen from Fig. 3.7. However, there is a price to pay for these positive impacts, which is that more water is consumed (Fig. 3.9) and, especially for the end of this century, this increase is expected to be substantial.

The overall effect of climate change on food production for the seven basins appears to be positive and one can ask whether adaptation strategies, as defined earlier, are required. The first reason is that although the impact of climate change is positive, it might be that with some adaptation strategies the positive impact can be even higher. An example is Volta Basin where more precipitation can be expected in the future, so irrigation can be intensified. A more important consideration to imple-

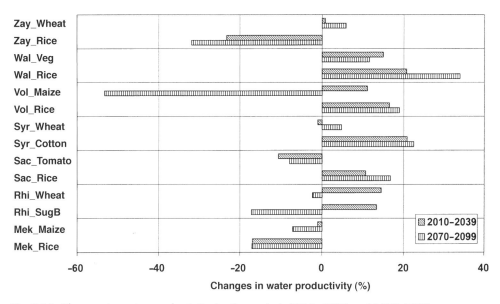

Fig. 3.10. Changes in water productivity for the periods 2010–2039 and 2070–2099 as compared with the baseline 1961–1990. Values for HadCM3 A2 climate change projections and business as usual strategy.

ment adaptation strategies at field scale is that at the basin scale changes will occur that will require responses at the field scale. An example is the Walawe Basin, where less water will become available for agriculture as a result of economic changes, which will result in lower water availability for agriculture. In other words, the adaptation strategies as defined will not lead automatically to higher yields.

Figure 3.11 depicts in summary what the impact of the defined adaptation strategies will be, compared to the business as usual strategy, for the period 2010–2039. In other words, the impact of climate change without any adaptation is reflected in Figs 3.7–3.11, and shows a positive trend, but what will happen if adaptation strategies are implemented? As mentioned before, some of the adaptation strategies will result in lower yields, compared to the business as usual situation, but these adaptations might be required due to basin-scale changes. However, most of the adaptation strategies analysed here will still generate higher yields in 2010–2039 compared to 1961–1990, but the business as usual approach performs better in half of the locations.

The following sections will provide for each basin a summary of the main results of the adaptation assessment at field scale. Figures 3.7–3.11 form the basis for these discussions, where the following points are important to consider.

- The baseline period (1961–1990) is used as reference for the future periods of 2010–2039 and 2070–2099.
- For each basin two crops are considered.
- For each basin two adaptation strategies are compared to the reference (business as usual).

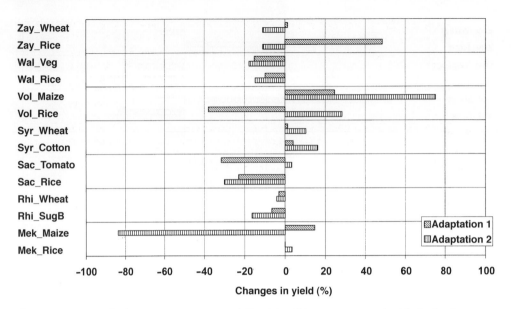

Fig. 3.11. Impact of adaptation strategies as defined in Table 3.2 compared with the business as usual option. The changes in yield show what the impact is if the adaptation strategy had been implemented for the period 2010–2039 (HadCM3, A2).

Mekong

For the baseline scenario, *rice* production will increase substantially in the future, especially according to the A2 scenario. The B2 scenario shows only a small increase for the distant future, but at the cost of a high variability. Water consumption is increasing by 100–150 mm. Since the Mekong receives a substantial amount of precipitation during the main rice season, the option to cultivate rice without irrigation was tested. Somewhat surprisingly, the predicted yield was almost similar and only a small increase in variation could be seen for the 1961–1990 period. Average annual precipitation during these years was about 1800 mm. Also for the future this non-irrigated rice seems to be an option, but variation in yield will increase substantially. Although the SWAP model runs on a daily basis, rainfall was only available on a monthly basis, so drier spells within a month were not included. Finally, what the impact of a shorter growing season would be was explored. Obviously, this assumes that crop varieties exist, or will be developed, with a short growing season of about 120 days. It appears that this option will result in the highest water productivity, but not in the highest crop yield in kg/ha. This might lead to a potential conflict of interest, where farmers are less interested in water productivity than water managers are.

Maize was introduced recently as crop to be planted after the flood season in December and January. Maize is not irrigated and, because of the residual water in the soil and supplementary irrigation, it is still possible to obtain a reasonable yield. The option to grow maize in the summer season (June–September) results in very low yields according to the model and many years will have no yield at all. This is somewhat surprising, since during the summer season average rainfall is around 1000 mm.

Table 3.2. Basic crop information and adaptation strategies considered for each basin and crop combination.

Basin	Crop	Adaptation strategy 1	Adaptation strategy 2
Mekong	Rice	Stop irrigation	Short season
	Maize	More irrigation	Summer season
Rhine	Sugarbeet	Lower groundwater	Short season
	Wheat	Lower groundwater	Spring season
Sacramento	Rice	Irrigation max. 900 mm	Deficit irrigation
	Tomato	Stop irrigation	Increased irrigation
Syr Darya	Cotton	Reduce soil evaporation	Increased irrigation
	Wheat	Reduce soil evaporation	Increased irrigation
Volta	Rice	Stop irrigation	Intensification
	Maize	Irrigation	Intensification
Walawe	Rice	Short season	Reduced irrigation
	Vegetables	Short season	Reduced irrigation
Zayandeh	Rice	Reduced salinity	Reduced irrigation
	Wheat	Reduced salinity	Increased salinity

A closer look at the model results indicates that crop yield is hampered by water surplus, and fields are completely saturated during prolonged periods. Since we have used the same soil type as for rice, except for the puddling layer, drainage is inadequate and therefore less favourable for maize. Obviously an enhanced drainage capacity would diminish these low yield potentials.

Rhine

In contrast to other basins and crops, *sugarbeet* yield is not highest in the period 2070–2099, but peaks in 2010–2039. The main reason is the projected lower precipitation, but especially the shift in precipitation patterns. Annual average precipitation will be reduced by less than 10% in 2070–2099, but rainfall in the sugarbeet growing season will reduce by 412, 386 and 231 mm/year for the periods 1961–1990, 2010–2039 and 2070–2099, respectively. At the same time, crop water requirements are higher due to increased temperatures.

In case of low groundwater tables, these effects are even more profound and sugarbeet yields will be lower as capillary rise, an important water source during dry periods, will be reduced. Also interesting is the enormous increase in expected yield variation for the 2010–2039 and 2070–2099 periods. A possible option to reduce these negative impacts of climate change might be to reduce the length of the growing period, providing suitable varieties exist. Although less water will be consumed and water productivity is similar, crop yields are lower and this option is not really solving the problem. In addition, prices for sugarbeet are related to sugar content of the beets, which is normally negatively affected by shorter seasons.

The *wheat* crop shows a pattern similar to sugarbeet, although the increase in yield for the period 2010–2039 is less profound (Fig. 3.7). The crop is also less sensitive to lower groundwater levels. Changing the growing season would not increase crop yields or water productivity (Fig. 3.11).

Sacramento

Rice yields are expected to increase by almost 50% for the A2 and 20% for the B2 scenario, mainly as a result of enhanced CO_2 levels (Fig. 3.7). However, to maintain this level of yield an increased irrigation application is necessary from an average 900 mm/year in the period 1961–1990, 1000 mm/year in 2010–2039, to 1150 mm/year in the period 2070–2099 for the A2 scenario. From a water productivity point of view, it is interesting that it is still worthwhile to do so (Fig. 3.10).

Since irrigation is a key issue in Sacramento, we have defined two adaptation strategies. One is that irrigation is restricted to use on average maximum 900 mm/year and the second is that irrigation is reduced by about 10% (deficit irrigation). The first option has no impact for the period 1961–1999, since this 900 mm is the actual amount used. For the near and distant future such a restriction would have a dramatic impact on yields (Fig. 3.11) and water productivity. From a basin perspective this implies that if no more water will become available for irrigation, some rice areas should be taken out of production. An overall reduction in rice irrigation of 10% will induce a substantial loss in production of about 30%. Note that with this strategy an increase in irrigation over time is still assumed, but less than during the baseline.

Tomato production provides the highest water productivity of all the cases considered in this study. It should be taken into account that we have defined WP here in terms of gross production, while production costs of tomato are known to be high. Somewhat remarkable is that variation in yield reduces over time for the baseline as well as for the increased irrigation strategy. However, variation is still high in comparison to the rice crop.

Increasing the amount of irrigation (adaptation strategy 2) is not boosting the production substantially. To stop irrigation completely (strategy 1) does not have a severe impact on yields for the period 1961–1990, but is dramatic for the future, especially for the period 2070–2099. For this period the projected rainfall decreases by almost 15% for the A2 and B2 scenario, while temperatures and thus crop water requirements are higher. The low impact of stopping irrigation for the 1961–1990 period can be explained by the favourable soil characteristics in terms of soil water-holding capacity.

Syr Darya

Trends in *cotton* yields for the future are positive. Yields will be higher and variation in yields will not change substantially. Reducing soil evaporation (adaptation strategy 1), for example by mulching the topsoil, will increase crop yields somewhat and also an increase in irrigation (adaptation strategy 2) will help to obtain higher yields. However, since we consider only field-scale issues here and since total water consumption will be higher in the future, the impact on basin-scale total production should be analysed as well. The option of trying to reduce soil evaporation is therefore very relevant in this respect. Comparing the A2 to the B2 scenario shows the same trends, but is less profound for B2.

Conclusions drawn for cotton are also valid for *wheat*. Higher yields in the future, more water consumed and reducing soil evaporation or increasing irrigation have a positive impact on yields.

Volta

The GCM outputs indicate that no major changes are expected in rainfall for the Volta Basin and climate change mainly will have an impact on temperature and CO_2 concentrations. The SWAP runs show that as a result of the latter, *rice* yields are expected to increase by about 45% and 30% for A2 and B2 scenarios, respectively (Fig. 3.7). Since rice is grown in the wet period, the result of growing rice only under rain-fed conditions was explored. Obviously, yields will be lower compared to irrigated conditions, but if we look at current and future rain-fed yields, no major changes in the mean are expected. However, variation in yields will be higher, and water productivity lower.

The adaptation strategy mentioned here as intensification (adaptation strategy 2) – which includes options such as improved crop variety, denser planting and shorter season – will increase crop yields. Water consumption will be lower and water productivity will be higher under this option. These positive effects are stronger under the A2 scenario than under the B2 scenario.

During the wet season *maize* is commonly produced under rain-fed conditions and therefore this practice is used as the baseline. A small increase in yields is predicted for the near future, but yields in the distant future are expected to reduce substantially by about 50% (Fig. 3.7). Introducing irrigation will increase yields and will reverse the lower yields expected for 2070–2099 to an increase. As with rice, crop intensification has been explored, but still without irrigation. Such intensification can increase mean yields, but since irrigation was not included, variation in yields will be much more profound.

Walawe

Rice production in Walawe will benefit substantially from climate change and increases in yield are expected to be about 45% and 20% at the end of this century for A2 and B2 scenarios, respectively. Only a small decrease in annual variation can be expected and also the increase in water consumption is less than 10%. Precipitation will also be about 10% higher, which will keep the water balance similar. However, it is expected that more water will be required for domestic, industrial and service-oriented activities, which will result in less water available to irrigation. One adaptation strategy was rice cultivation with less irrigation water (adaptation strategy 2) and results indicate that yields will be lower even with a small decrease in irrigation. A detailed study on this adaptation strategy, where field and basin water resources were linked, showed a strong reduction in yields under lower irrigation intensities.

The other adaptation strategy (adaptation strategy 1) explored whether a shorter season would be worth implementing. According to the simulations, such an approach is indeed an option and, although rice yields will be somewhat lower, water productivity will increase. It was assumed here that the appropriate rice variety would exist.

The second crop considered is referred to as *vegetables*. Although a wide variety of vegetables exist, each with its own characteristics, we have used here a kind of reference vegetable crop. One of the most striking results is that an enormous increase in variation in yields is to be expected in the future. A closer look at the model results shows that this variation was mainly due to water surplus instead of water shortage. The soils considered are not well drained, which is an advantage for rice, and combined with the expected increase in variation in rain will result in soil moisture conditions that are too wet for vegetables during some years. The two adaptation strategies, less irrigation and short season, have the same impact on vegetables as on rice: yields will be lower but also variation in yields will reduce. The short season option is also beneficial in terms of water productivity.

Zayandeh

Under baseline conditions, *rice* yields will be much lower in the future. Model output shows that salinity is the major reason for this. Although salinity levels in irrigation water are low (0.3 dS/m), this is already sufficient to have salt accumulation in the soil so that crop growth will be hampered and yields will be lower. It might be that the ECHAM4 projections with somewhat more rainfall in the winter would boost leaching and the impact of salinity will be less.

The adaptation strategy in which we assumed that salinity in irrigation water is zero shows that salt is one of the major threats to agriculture in Zayandeh. Rice cultivation in particular is very sensitive to salinity. The second adaptation strategy, where irrigation will be somewhat lower, shows a major drop in yields. Detailed analysis of the impact on irrigation supply and salinity can be found elsewhere (Droogers and Torabi, 2002).

For *wheat*, the impact of salinity was explored further by including an increased as well as a decreased salinity level. Results show that although salt has an impact on expected yields, this is much lower than for rice. In terms of future projections, climate change will have a positive impact on wheat production in Zayandeh.

It should be emphasized again that results as presented are valid for field-scale analysis, without taking account of basin water resources. For Zayandeh this is essential, since total consumed water will increase for rice as well as for wheat at field level and whether this water is available should be explored at basin level. If not, two options are available to adapt: reduced irrigation depth or reduced cropped area.

Conclusions

One of the most striking conclusions of this study is that the overall impact of climate change on crop yields is positive. Figure 3.7 illustrates that in the business as usual strategy expected yields are higher for all basins except for one basin–crop combination in the period 2010–2039 and for two basin–crop combinations in the period 2070–2099. Also variation in crop yields is going down for more than half of the basin–crop combinations (Fig. 3.8), despite the increase in climate variability in the HadCM3 projections. However, there is a price to pay for these positive impacts and that is that more water is consumed (Fig. 3.9) and, especially for the end of this century, this increase is expected to be substantial.

Water productivity (Fig. 3.10) shows a mixed picture, with some basin–crop combinations having 10–20% higher values and others 50% lower values. It is important that water productivity values are a function of market prices, which we have assumed constant here. Although this is highly unlikely, it provides good insight into the combined yield and water consumption processes.

Future crop yields are potentially higher as a result of enhanced CO_2 levels. At a basin scale it has to be evaluated whether the increased amount of water required for the crops is available. As mentioned in the introduction, this chapter only describes the field scale component of the ADAPT project and linkages to the basin scale will be described in the other chapters. The adaptation strategies as explored here can be used in basin-scale studies to evaluate alternative options for dealing with increased water consumption.

One of the dominant factors in the analysis in this chapter is the impact of elevated CO_2 levels on crop production. This impact, and especially the long-term impacts and feedback, is still under debate. However, results of numerous experimental studies as described in the methods section indicate that a doubling in CO_2 levels can indeed induce higher yields by up to 50%.

We employed the relation of Doorenbos and Kassam (1979) to derive relative crop yield as a function of relative transpiration. In order to apply this methodology, the potential crop growth in a certain weather year had to be estimated from the potential transpiration (Droogers and van Dam, 2004). The adopted methodology has the tendency to overestimate the variation of crop yields in a sequence of years. Further studies with simple but reliable crop growth routines are required.

The increase in variation in crop yields is not as dramatic as expected. One of the most important reasons for this is the way irrigation scheduling was included in the simulations. The irrigation scheduling option in SWAP implies that if a crop experienced stress beyond a defined threshold value, irrigation was considered to take place. The consequence is that during dry years, through either low rainfall or high temperatures, more irrigation would be applied than during wet years. This is only a realistic assumption when sufficient storage capacity exists in a basin. This scheduled irrigation approach guarantees that irrigation timing is always optimal, ensuring that results are not affected by this timing but only by the total water applied.

References

Adams, R.M., Hurd, B.H. and Reilly, J. (1999) Agriculture and global climate change: a review of impacts to U.S. agricultural resources. Pew Center on Global Climate Change. Available at: http://www.pewclimate.org/projects/env_agriculture.cfm

Bazzaz, F. and Sombroek, W. (1996) *Global Change and Agricultural Production*. Food and Agriculture Organization of the United Nations (FAO), John Wiley & Sons, New York.

CERUDEB (2002) A small farmer crop micro-insurance scheme in Uganda. Centenary Rural Development Bank Uganda. Available at: http://www.mcc.or.ug/downloads/agriculture.doc

Chen, S. and Ravallion, M. (2000) How did the world's poorest fare in the 1990s? World Bank Policy Research Working Paper No. 2409. World Bank, Washington, DC.

CSCDGC (2002) Plant growth data. Center for the Study of Carbon Dioxide and Global Change, Tempe, Arizona. Available at: http://www.co2science.org

Doorenbos, J. and Kassam, A.H. (1979) Yield response to water. Irrigation and drainage paper 33. Food and Agriculture Organization of the United Nations, Rome.

Droogers, P. and Torabi, M. (2002) Field scale scenarios for water and salinity management by simulation modeling. IAERI-IWMI Research Reports 12. Colombo, Sri Lanka.

Droogers, P. and van Dam, J.C. (2004) Field scale adaptation strategies to climate change to sustain food security: a modeling approach across seven contrasting basins. IWMI Working Paper. International Water Management Institute, Sri Lanka.

EC (2003) Facts and figures on European agriculture. Available at: http://europa.eu.int/comm/agriculture/publi/index_en.htm

FAO (2002a) *Food Insecurity: When People Must Live with Hunger and Fear Starvation. The State of 2002.* Food and Agriculture Organization of the United Nations, Rome.

FAO (2002b) *World Agriculture: Towards 2015/2030.* Food and Agriculture Organization of the United Nations, Rome.

FAO (2003a) International commodity prices. Food and Agriculture Organization of the United Nations, Rome. Available at: http://apps2.fao.org/ciwpsystem/wpnote-e.htm and http://apps2.fao.org/ciwpsystem/ciwp_q-e.htm

FAO (2003b) Global information system of water and agriculture. Food and Agriculture Organization of the United Nations, Rome. Available at: http://www.fao.org/waicent/faoinfo/agricult/agl/aglw/aquastat/main/ index.stm

FFTC (2003) Research for the production of major food crops in Vietnam. Food and Fertilizer Technology Center. Available at: http://www.agnet.org/library/article/bc46015.html

Fischer, G., Shah, M., van Velthuizen, H. and Nachtergaele, F.O. (2001) *Executive Summary Report: Global Agro-ecological Assessment for Agriculture in the 21st Century.* International Institute for Applied Systems Analysis, Laxenburg.

Hartz, T.K. (2003) Processing tomato production in California. Vegetable Research and Information Center. Available at: http://vric.ucdavis.edu/ selectnewcrop.tomato.htm# production

IRS (2003) Instituut voor rationele suikerproductie (in Dutch). Available at: http://www.irs.nl/irs_betatip.htm

Koyama, O. (1998) Projecting the future world food situation. Japan International Research Center for Agricultural Sciences Newsletter 15. Available at: http://ss.jircas.affrc.go.jp/kanko/newsletter/nl1998/no.15/04koyamc.htm

Kukla, G. and Karl, T.R. (1993) Nighttime warming and the greenhouse effect. *Environmental Science and Technology* 27, 1468–1474.

Lindeman, M. (2003) Crop production in the cotton region of the former Soviet Union. Available at: http://www.fas.usda.gov/remote/soviet/country_page/fsuasia_text.htm

Mendelsohn, R. and Dinar, A. (1999) Climate change, agriculture, and developing countries: does adaptation matter? *The World Bank Research Observer,* 14(2), 277–293.

Parry, M.L. and Rosenzweig, C. (1993) Food supply and risk of hunger. *Lancet* 342, 1345–1347.

Rosegrant, M.W., Ximing, C. and Cline, S.A. (2002) *Water and Food to 2025: Policy Responses to the Threat of Scarcity.* International Food Policy Research Institute, Washington, DC.

Rosenzweig, C., Parry, M.L., Fischer, G. and Frohberg, K. (1993) Climate change and world food supply. University of Oxford, Environmental Change Unit, Research Report No. 3.

Rötter, R.P. and van Diepen, C.A. (1994) Rhine basin study: land use projections based on biophysical and socio-economic analyses. Volume 2. Climate change impact on crop yield potentials and water use. DLO Winand Staring Centre, Report 85, Wageningen.

Seckler, D., Barker, R. and Amarasinghe, U. (1999) Water scarcity in the twenty-first century. *Water Resources Development* 15, 29–42.

Shiklomanov, I.A. (2003) *World Water Resources at the Beginning of the 21st Century.* Cambridge University Press. Summary available at: http://webworld.unesco.org/water/ihp/db/shiklomanov/index.shtml

Smit, B. and Skinner, M. (2002) Adaptation options in agriculture to climate change: a typology. *Mitigation and Adaptation Strategies for Global Change* 7, 85–114.

Tobey, J., Reilley, J. and Kane, S. (1992) Economic implications of global climate change for world agriculture. *Journal of Agriculture and Resource Economics* 17, 195–204.

UNEP (1998) *Handbook on Methods for Climate Impact Assessment and Adaptation Strategies.* Feenstra, J.F., Burton, I., Smith, J.B. and Tol, R.S.J. (eds) Institute for Environmental Studies, Vrije Universiteit, Amsterdam.

van Dam, J.C., Huygen, J., Wesseling, J.G., Feddes, R.A., Kabat, P., van Walsum, P.E.V., Groenendijk, P. and van Diepen, C.A. (1997) Theory of SWAP version 2.0. Technical Document 45. Wageningen Agricultural University and DLO Winand Staring Centre, Wageningen.

van Diepen, C.A., Wolf, J., van Keulen, H. and Rappoldt, C. (1989) WOFOST: a simulation model of crop production. *Soil Use and Management* 5, 16–25.

Vörösmarty, C.J., Green, P., Salisbury, J. and Lammers, R.B. (2000) Global water resources: vulnerability from climate change and population growth. *Science* 289, 284–288.

Wand, S.J.E., Midgley, G.F., Jones, M.F. and Curtis, P.S. (1999) Responses of wild C4 and C3 grass (*Poaceae*) species to elevated atmospheric CO_2 concentration: a meta-analytic test of current theories and perceptions. *Global Change Biology* 5, 723–741.

Wösten, J.H.M., Lilly, A., Nemes, A. and Le Bas, C. (1998) Using existing soil data to derive hydraulic parameters for simulation models in environmental studies and in land use planning. DLO Winand Staring Centre, Report 156, Wageningen.

4 Water for the Environment: Exploring Adaptations to Climate Change in River Basins

RALPH LASAGE AND JEROEN AERTS

Institute for Environmental Studies, Vrije Universiteit Amsterdam, Amsterdam, The Netherlands

Introduction

Research shows that changes in water quality and water quantity may affect sustainability of aquatic and terrestrial ecosystems. For centuries, natural conditions in rivers have been affected and human activity has reduced the capacity of water resources to support ecosystems and biodiversity (Covich *et al.*, 1997; Poff *et al.*, 2002). Ecosystems, however, have an important role in the hydrological cycle, as they have natural cleaning capacity and reduce the concentration of many (organic) pollutants in water. Ecosystems also help to reduce extremes in runoff through their capacity to store water and play an important role in supplying food, fibre, wood and medicine, and of course support the existence of many species (Gilbert and Janssen, 1998). Since the Dublin Statement (1992) on water and sustainable development, the importance of water for the environment and the environment for water is increasingly taken into account. To date, however, the role of ecosystem security in water management and in the development of adaptive strategies has been minor, and should deserve greater attention (Kabat and van Schaik, 2003).

Apart from pressures as populations increase and land use changes, climate change (CC) will affect the capacity of water resources to support ecosystems. Important factors are changes in precipitation, temperature, runoff and ecological processes (Band *et al.*, 1996; Lettenmaier *et al.*, 1999; IPCC, 2001). Climate change has the potential to increase the impact of other pressures that are currently affecting the environment and it is irreversible. Studies on water and ecosystems have not addressed this issue adequately, since the focus is either at the species scale or at the higher – global – ecosystem scale. Studies at the intermediate – basin – scale are lacking and integral impact or adaptation studies to look both at CC impacts to the environment and other water uses (food, industry, etc.) are scarce. Only a few studies have a more integrated approach across different scales. White *et al.* (1999), for example, show how vegetation responds to CC at different scales.

The ADAPT study has focused on the integrity of ecosystems under climate change and assessed different adaptation options to achieve sustainability. In this project, ecosystems are treated as one of the users of water in a catchment, apart from agriculture, industry and domestic use (Aerts *et al.*, 2003). Consequently, ADAPT pays attention to different adaptation measures ranging from socio-economic to engineering measures, as recommended by working group II in the IPCC Third Assessment Report (TAR) (Klein and MacIver, 1999). As agriculture, industry and domestic use often compete with environmental users, ADAPT addresses the trade-off between the different water users in the basin and the tension that is raised by this trade-off (e.g. Doll *et al.*, 2003).

This chapter focuses on environmental aspects of adaptation to climate change and describes how to develop and evaluate environmental adaptations to climate change for river systems. First, the possible effects of climate change on the environment, through changes in water quantity and quality are addressed. The possible environmental effects related to water and climate change are briefly described on a global and regional scale. Then the AMR framework (see Chapter 1) is used to specifically develop and evaluate adaptations with respect to environmental security. The method will be illustrated using two case studies. One case study is located in Central Asia (the Syr Darya Basin, Chapter 5) and one case study is situated in Western Europe (Rhine Basin, Chapter 7).

Environmental Issues Related to Water and Climate Change

This section first discusses the status of water resources and environment at the global scale, to get a grasp of the overall distribution of hot spots. Then we zoom in to the regional and local scale, to see through which processes climate change affects the environment, since at these scales basins are managed and potential adaptation measures are taken to meliorate the effects of CC.

Global water resources and ecosystems

In Fig. 4.1, the long-term average mean annual runoff (MAR) of rivers is shown. This is calculated using a global runoff model (Smakhtin, V.U., Revenga, C. and Döll, P., unpublished). Between regions there is a large difference in runoff. The dark areas show high runoff and the light areas low runoff. As can be expected, high yearly runoff occurs in the tropics and the yearly runoff in the desert regions is very low. The centre of North America, Australia and central Asia are places in the world where the water resources are scarce. These are the potential hotspots when CC leads to a reduction in runoff. Ecosystems need a certain amount of water to subsist and this amount varies between regions and ecosystems. For the majority of the rivers in the world there is no recommendation on the environmental water requirement. To estimate the amount of water that ecosystems need worldwide, a modelling study is performed (Smakhtin, V.U., Revenga, C. and Döll, P., unpublished). This study used the environmental water requirement as the definition of the amount of water needed to sustain ecosystems.

The environmental water requirement (EWR) is calculated by adding the

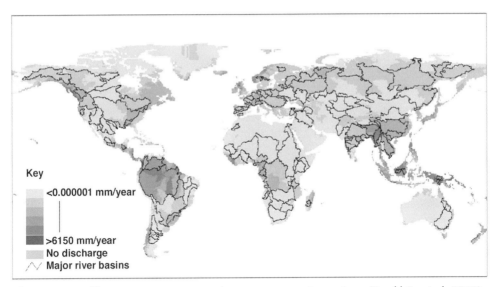

Fig. 4.1. Map of long-term average annual water resources in mm/year (Smakhtin *et al.*, 2003).

environmental low flow requirement (LFR) to the environmental high flow requirement (HFR). The LFR of a river is defined as the average runoff that is exceeded 90% of the time throughout the year, in other words, during 10% of the year runoff is lower. This information can be extracted from a flow duration curve of a river as shown in Fig. 4.4. The HFR is also defined as the discharge exceeded during a certain percentage of time a year. This percentage varies between different types of rivers. In a highly variable regime, the impact of changes in high flow on the system will be more severe than in a more stable system. For rivers with a highly variable flow, the HFR is equal to the runoff that is exceeded 20% of the time on average throughout the year. For rivers with a very stable flow, the HFR is considered to be zero.

Defining the EWR as above is a first step of estimating the health of ecosystems in relation to water availability. It is a very general method, and it does not take into account a desired state of ecosystems in the basin or an environmental management class in which an ecosystem needs to be maintained. That was not feasible in this first global-scale assessment of environmental water requirements (Smakhtin, V.U., Revenga, C. and Döll, P., unpublished).

The global EWR projections and calculations show that globally 20–50% of the mean annual river runoff is needed to sustain ecosystems in the current status. The light coloured areas in Fig. 4.2 represent the basins where a small proportion of the runoff is needed for sustaining the ecosystems and in the dark areas up to 50% of the runoff is needed for sustaining the ecosystems. These estimations could be on the low side because of the assumptions made in the model (Doll *et al.*, 2003). For instance, the sensitivity and importance of the (aquatic) ecosystem were not taken into account. When an ecosystem is very important, a larger proportion of the water could be allocated to it.

When ecosystems are regarded as a water user, a large proportion of the water resources should be allocated to them. Hence, knowing that human water withdrawals

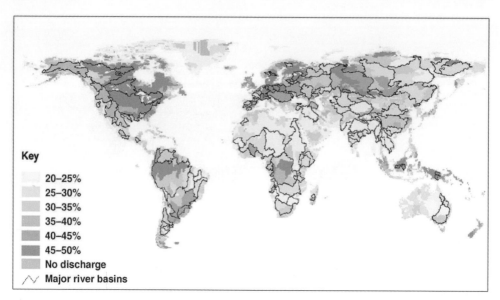

Fig. 4.2. Environmental water requirements for sustaining ecosystems as percentage of total runoff (Smakhtin *et al.*, 2003).

increase and that climate change may decrease water availability, further strains on the water resources in the basin can be expected.

By combining EWR and water availability, a Water Stress Indicator (WSI) can be calculated as shown in Fig. 4.3. The WSI is the water withdrawal (the EWR) as a proportion of water availability. This is shown in Fig. 4.3, where water stress in basins is presented, accounting for the environmental water requirements as shown in Fig. 4.2. The basins with a WSI higher than 0.7 are basins where there is not enough water to meet the environmental water requirements. Figures 4.2 and 4.3 are based on values under the current climate. It is expected these figures will change under influence of climate change and in the future more problems will exist between minimal environmental requirements and human use of water and the available water. In the marked basins, the water stress increases in comparison to the traditional water stress indicators, water withdrawal as a proportion of the average total water resource. In the dark areas the current WSI is very high and under CC this stress will probably increase, for instance around the Mediterranean where the precipitation is expected to decrease under CC (IPCC, 2001) and mid- and west North America.

Change in mean annual water availability is not the only quantitative indicator that measures stress to ecosystems. One can also look more specifically at low flows and high flows, since ecosystems are largely influenced by the occurrence of periods of low flows and high flows. These features can be similarly used as indicators for presenting environmental stress in river basins and, especially for low flow hydrology, much work has been done (Caruso, 2001; Smakhtin, 2001). Low flows can vary over the years and are influenced by precipitation, snow melt and anthropogenic impacts, like groundwater abstraction, artificial drainage, changes in vegetation, deforestation, discharge from storage and water extraction for industry, agriculture and municipal use. They are

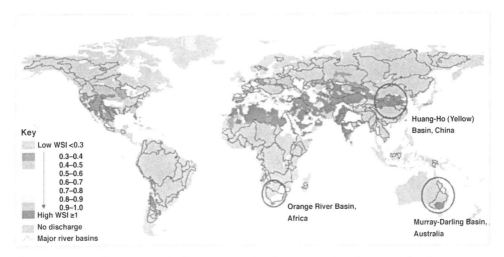

Fig. 4.3. A map of water stress indicators that takes into account environmental water requirements in river basins. The circles indicate case studies used by Smakhtin *et al.* (2003).

important since ecosystems need a minimum discharge to ensure fish passage, maintain certain temperature levels, habitat maintenance, sedimentation control, etc. Some ecosystems also need a high flow for a yearly flooding, like floodplains ecosystems.

A so-called flow duration curve shows how to address changes in low flow and high flow (Fig. 4.4). This curve displays river discharge at a certain point along the river, in relation to the percentage of time a certain discharge has occurred. In Fig. 4.4 there are two curves of a river; the continuous line represents the current situation; the interrupted line shows the future situation under climate change. The discharge that occurs under the current circumstances for 100% of time is the value on the right side of the graph. In this case the minimum discharge is approximately 1600 m^3/s. Towards the left of the graph, the probability that the discharge is exceeded becomes lower, in other words it occurs less frequently during the year. For instance, the chance that a discharge of 3500 m^3/s is exceeded is 35%. The probability of exceeding a discharge decreases with an increase in discharge. This continues until the probability is zero, which is the maximum discharge of the river. The maximum discharge of this river is 8000 m^3/s. From the graph, low and high flow discharges can be derived. When the minimum and maximum discharge requirements for the ecosystems in the basin are known, these can be compared with the flow duration curve of the river. The interrupted line in Fig. 4.4, the future situation under climate change, shows a shift in the minimum discharge of the river to 500 m^3/s. When, for instance, the minimum flow requirement of the ecosystem is 1000 m^3/s, this is no longer met in the future. This information is important to a water manager of the basin. The management of reservoirs can be changed to increase the instream of water to the river through reservoir releases during extreme low flow periods. Hence, low flow requirements can be met in the basin. If the changes in low flow are allowed, these can negatively influence the habitat dynamics of a system and in the long term can change the ecosystems (Dakova *et al.*, 2000).

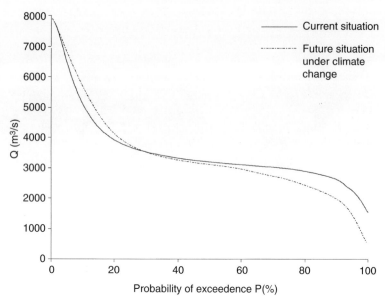

Fig. 4.4. Example of a flow duration curve.

Regional water resources and aquatic ecosystems

Global changes in water resources described earlier in this chapter can be refined on a basin scale. A basin can be regarded as a natural hydrological unit, which contains the total water system within its borders and has different users that are all dependent on the same water resource. There is no exchange with other basins or resources, so the basin can be seen as a whole. On this regional scale it is useful to make a distinction between aquatic and terrestrial ecosystems, because there is a difference between the issues and expected climate change effects they have to cope with. The impacts described in this paragraph are possible effects, which obviously vary between different regions and ecosystems.

Aquatic ecosystems
Climate change is an additional threat to aquatic ecosystems and it interacts with other human-caused stresses like pollution and fragmentation of landscape. Aquatic systems are increasingly disconnected and isolated, making adjustment to the changes through animal and plant dispersal difficult. This certainly applies to the riverine and coastal ecosystems (Poff *et al.*, 2002). For aquatic ecosystems, a distinction is made between freshwater systems and saltwater systems. Lakes, streams, wetlands and rivers are part of the freshwater ecosystems; saltwater ecosystems contain the sea, corals and estuaries.

It is expected that freshwater lakes will be influenced through an increase in air and water temperature. Many lake systems as well as rivers and streams are shallow, well mixed, meaning they exchange heat and oxygen easily with the atmosphere. Under rising air temperatures in the coming century, the temperature of the rivers, streams and lakes will also increase. The average global surface temperature is expected to increase by 1.4–5.8°C by 2100 relative to 1990 (IPCC, 2001). If, for example, the water temperature increased by 4°C in a region, the present day ecosystems would have to migrate 680 km to higher latitudes, in order to maintain the same thermal regime (Sweeny *et al.*, 1992). This leads to a reduction in cold water habitats and might cause a decrease in cold water species like trout and salmon and an increase in warm water species like bass (Mohseni *et al.*, 2003). For species in isolated systems, like lakes and wetlands without corridors, the rise in temperature will mean their demise. A study on the effects of CC on lake trout in Alaskan lakes concludes that these species might not survive under future climate circumstances and this will lead to major food-web changes in Arctic lakes (McDonald *et al.*, 1996). Besides the disappearance of invertebrate species from areas, the increase in water temperature has the potential to lead to an increase in algae blooms and pests like botulism (Poff *et al.*, 2002).

The summer of 2003 was very dry in Northwest Europe, the catchment of the Rhine River. During the first half of the year less precipitation fell and this affected the water level of the Rhine. In August, the lowest levels ever were measured. These low levels led to problems with transport along the river and with water temperature, because the drought coincided with record-breaking temperatures. The summer of 2003 was the warmest summer in 500 years in Europe, according to the University of Bern. Fish in the river were affected by these changes. The eel, for instance, migrates during the summer from upstream areas to the sea. In 2003, many were infected by a bacterial plague, which was a result of the high water temperature. The high temperature of the water, sometimes higher than 30°C, caused them severe stress. Their poor condition and the low water level led to the death of tens of thousands of eels, as many of them were injured by the screw propeller of boats and many died as a result of their infections. At the same time, the governments in the Netherlands and Germany gave special permission to energy plants to release cooling water into the Rhine above the legislative temperature of 30°C, which further increased the heat stress on the system.

The sources of many rivers lie in mountainous regions, for example the Alps for the Rhine in Europe and the Tian Shan mountains for the Syr Darya in Central Asia. Changes in hydrology in the upper part of the basin may have implications downstream of the river (Lettenmaier *et al.*, 1999). Aquatic ecosystems and riparian terrestrial ecosystems can be affected indirectly by changes in the mountains induced by CC. It is expected that a greater proportion of precipitation will fall as rain instead of snow and that the melting season will shift to an earlier start in the year. The summer base flow and (spring) peak flow will change as a result (Frederick and Gleick, 1999). Many species, aquatic and riparian, are sensitive to changes in the frequency, duration and timing of extreme events like floods and droughts. Some of them have adapted their reproductive strategies to avoid or to take advantage of the spring floods. An example is the salmon, which depends on the spring floods for its migration to spawning grounds (Poff *et al.*, 2002). Changes in precipitation can lead to a reduction in inflow to lakes. When this decrease in inflow lasts for a longer time, the level of the lakes will drop and run the risk of drying out like the Aral Sea. The opposite is also possible. The level of the Caspian Sea, for example, has been rising for the last few decades, but this fluctuation is probably not driven by CC.

Wetlands are adjusted to their local circumstances with periodical floods and droughts. Water dynamics make these areas very productive. Fluctuations in the current water regime could change the timing and duration of the droughts and floods, and will affect the ecosystem. If the depth or duration of flooding changes, the area will become habitable for other species that are able to cope with the new circumstances. This will be terrestrial species when flooding decreases or aquatic species when flooding increases. Coastal wetlands and mangroves also depend on the inflow of freshwater from the rivers. The height of sea level is another factor of influence on coastal ecosystems. A reduction of inflow of fresh water, or a rise in sea level, will lead to an increase in salinity level. If these changes last for a longer period, species may not endure the strain of the increased salinity level and they will be replaced by other species that can cope with these levels (Blasco *et al.*, 1996; Ellison *et al.*, 2000). If the sea level rise is too fast, ecosystems cannot keep pace with this rise and they will drown. This could be prevented in regions where the supply of sediment is enough to keep up with sea level rise. This is not likely to be the case in many regions, because of the anthropogenic alterations of the rivers that prevent large quantities of sediment being transported downstream. With the disappearance of mangrove forests, the natural defence against the sea disappears. The shoreline will be vulnerable to erosion and the coastal zones will be impacted.

Terrestrial ecosystems and biomes

An ecosystem is a system of interactions of a community of organisms with its surroundings. The largest fundamental region of an ecosystem is the biome. A biome consists of similar plants and animals that can be distinguished from each other by vegetation and climate. Some examples of biomes are grasslands, tundra, forests and tropical rainforests (Bryant, 1997). A biome includes both aquatic and terrestrial ecosystems. Two laws control the boundaries of climatically determined biomes. These are the law of the minimum and the law of the maximum (Bryant, 1997). The law of the minimum states that plants have a minimum requirement for growth and reproduction controlled by a range of factors, like nutrients, availability of water, sunlight and temperature. The factor that is least available has the greatest effect on plant growth, so that small changes in that parameter can have a profound influence on a plant's survival. In Fig. 4.5, the distribution of biome types depending on latitude, temperature and precipitation is shown. For instance, the disappearance of tropical forest in favour of savannas as the result of a decrease in precipitation is an example of change through not meeting the minimum requirement limit for the rain forest (Zheng and Lei, 1999). The law of the maximum states that too much of a certain factor also limits the area where a plant can survive; for instance, the replacement of wetland species by aquatic plant species when the water level rises in an area, or the replacement of one mangrove species by another, when the concentration of salt increases and exceeds a certain threshold (Ellison *et al.*, 2000). As a result of these laws and other environmental factors, biomes show up spatially as distinct zones of vegetation, mainly spaced parallel to the temperature gradient that exists either latitudinally between the equator and the poles, or with elevation on mountains.

Climate change affects the distribution of biomes through both of these laws. An increase in temperature exceeds the maximum requirements for some species, which will disappear from that place. For other species the rise in temperature could make

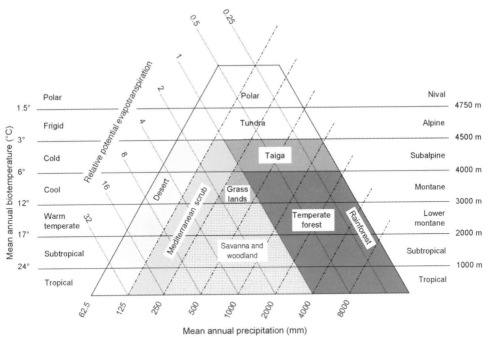

Fig. 4.5. Classification of biome types based upon climatic classification of Holdridge (1947).

certain places more suitable for habitation and they will migrate to these places. The same effects can be expected for changes in precipitation (Bryant, 1997).

Important drivers for changes in terrestrial ecosystems are migration, changes in fire regime and changes in water availability. How these changes affect the terrestrial biomes situated in river basins, forests, grasslands, plains and mountains is now described.

Migration of species

As stated above, an important issue related to climate change is migration of species. The expected migration caused by changes in temperature is a poleward migration and hence (sub-)tropical biomes will migrate from the equator to the direction of the poles. The migration is a direct result of the law of maximum and minimum that determines the reach of a biome. If the rate of the migration of biotopes is too high, the species which normally live in this biotope can not keep pace. It is possible these species will become extinct, but these impacts seem to be small (Morgan *et al.*, 2001). An analysis for species of birds, butterflies and herbs resulted in a mean shift of the range boundaries of 6.4 km per decade poleward, or 6.4 m per decade upward (Parmesan and Yohe, 2003). Another outcome of this study was the mean shift to an earlier spring of 2.3 days per decade. In some regions the growth rate of trees is decreasing (Barber *et al.*, 2000) and in other regions the changed circumstances have led to the replacement of one tree species by another (Allen and Breshears, 1998). In history, similar shifts and changes in vegetation have occurred during the alternation between glacial periods and deglaciation periods.

Forest fires

In temperate forests, one of the most important and much studied effects of climate change is expected to be a change in forest fire regimes, which eventually can lead to species migration. The increase in temperature and changes in the rainfall pattern as a result of CC are expected to provoke more prolonged droughts, which can led to an increase in forest fires. Also, the increase in the number of thunderstorms is expected to lead to more lighting strikes, which will cause more wild fires. These fires will influence the hydrology of a basin through the alteration of the age distribution of the trees in the forest (Valeo *et al.*, 2003) and possibly through the replacement species. This could be a slow growing species by a fast growing species, which profits from the reccurring fires.

Changes in regional water availability may affect ecosystems. The study of White *et al.* (1999), for example, implies a loss in area of rainforest in the Amazon Basin in the 21st century, in favour of savanna. This is the result of the expected increase in temperatures and decreasing precipitation. The same effect is predicted for tropical grasslands, which will be turned into deserts as a result of decreases in precipitation.

The IPCC further identifies other possible effects of CC on terrestrial ecosystems, which are the following: the positive effect of increased CO_2 concentrations on the growth and productivity of vegetation, changes in moisture content of soils, changes in nutrient cycling and changes in 'El Niño' events.

Framework for Evaluation of Adaptation Measures

The pressures on ecosystems in combination with the expected effects of CC, as described above, make it necessary for a water manager to develop adaptation strategies to maintain the quality of the ecosystems. In this research the emphasis lies on water-related changes and impacts, so the issues are approached through the perspective of a water manager.

Enhancing the quality of the environment is one of the goals of the water manager, as described in Chapter 1. The other goals are to enhance human welfare, enhance industrial capacity and to enhance food capacity and security. For reaching these goals, the AMR framework (Chapter 1) is developed to structure the development and evaluation of adaptation measures that address policy objectives and the physical state of the water resources system. This section describes the framework, with the emphasis on the environment.

To link the objectives to water quantity and quality, a set of indicators is used, as shown in Fig. 1.10 (p. 18). On the right, state indicators (SIs) represent the state of the water resources system. On the left, objectives are refined in decision indicators (DIs). Note that the objective 'Quality of nature' is elaborated in two intermediate objectives, 'habitat' and 'water quality'. This elaboration is based on a literature survey of concepts such as ecosystem health, ecological integrity and biodiversity (Lorenz, 1999). The quality of nature in a river system depends heavily on habitat (expressed in structure, area and biodiversity) and water quality. The DI describes the state of the (intermediate) objectives. Values of the DIs reflect whether the goals of a decision maker are reached.

There are many different DIs and the managers and decision makers involved have to choose the ones that are representative for goals in their basin. In Fig. 1.10, the DIs that are used in the ADAPT project are described. They enable the evaluation of the performance of measures taken by the water manager.

The state indicators characterize the state of the water resources system (WRS) of a basin in terms of water quantity and quality. Out of the effects of CC on the environment, as described earlier, indicators can be deduced. The environmental SIs based on these effects are: number of floods (environmental high flow), droughts (environmental low flow), temperature of the water and concentration of NaCl and BOD. During the project, other SIs suitable for describing the state of the WRS were developed. These are concentration of PCB, N and P and annual discharge. These indicators can be found in Fig. 1.10 and are described in Box 4.1.

The state and decision indicators are related to each other. If a state indicator changes (thus the water system of a river changes), the decision indicator will also change, and hence can imply that a certain objective cannot be met. Or, the other way around. When, for example, the goal is to bring a certain fish species like the salmon back into the river, the following steps have to be taken. First the current state of the water resources system has to be assessed. For instance, the concentration of fertilizer in the water and the absence of spring floods are the limiting factors for the return of the salmon. These limiting factors can be found through the indicators we assigned to measure the state of the WRS. The next step is to design and implement measures to improve the status of water resources. Such measures could be to reduce cropland near the river and to change reservoir management by allowing more spring floods. The effectiveness of such measure is evaluated through the changing values of the DIs.

With this framework it is possible to show the effects of interventions in the water resources system. It is, however, important to note that only the indicators that are applicable to the regional circumstances in a river basin can be chosen.

Environmental Adaptation Measures

The Syr Darya and Rhine Basins are used for demonstrating the AMR framework for evaluating adaptation strategies on the basis of environmental indicators (see Chapter 1). The main environmental issues and policy goals of the basins are briefly described and expressed in terms of indicators. This information is based on the reports of the basin studies conducted during the ADAPT project.

For the Syr Darya, the largest environmental problems correlate with the main activity in the basin, agriculture, and are partly caused by the way the water resources are managed (Savoskul *et al.*, 2003). Agriculture pollutes the water with pesticides, herbicides, insecticides and mineral fertilizers. Furthermore, the agricultural use of water is so high that the inflow to the Aral Sea has decreased drastically and the sea level is dropping, which has led to the collapse of ecosystems in the Aral Sea region since the 1960s. Besides the current pressures on the system, climate change will also impose additional pressures on the system in the future. These pressures are called exogenous influences in the framework, because they cannot be influenced by actions in the basin. Other exogenous influences are population growth and economic growth. For the Syr Darya it is expected that under climate change the runoff peak in spring will

Box 4.1. Ecosystem-related environmental indicators. The indicators fit into the AMR methodology described in Chapter 1.

WATER QUALITY

State indicators

[ ]BOD
Biochemical oxygen demand concentration. This measures the amount of oxygen required or consumed for the microbiological decomposition of organic material. It indicates how much organic matter is discharged by human activities. The BOD is lowered when sewage treatment plants are constructed. High BOD levels generally indicate low water quality.

[ ]NaCl
The concentration of NaCl in the river near the mouth will increase when the sea level rises and when the fresh water flow of the river decreases. Also the overuse of the freshwater resources by agriculture can lead to salinization. It may disturb the freshwater habitats, reduce food production and endanger surface drinking-water facilities. It is best to choose a representative station downstream in an area where salinization occurs. It is possible to value on the ordinal scale when no conductivity data are available.

[ ]PCB
This indicator measures the concentration of pollutants like pesticides, herbicides, insecticides and industrial by-products. PCB is an example of these pollutants and is a chronic pollutant. Concentrations of such contaminants in fish give an indication of the concentration of these pollutants in water and aquatic soil. These pollutants are obviously bad for human health. The most widespread pollutants in the ADAPT basins are PCB, DDT, mercury, lead, dioxin, other heavy metals and pesticides. These chemicals bind on fat and accumulate in the food chain. Other fish species can be used as indicators. For instance, for the Rhine the 'Eel' species is used.

[ ]Fertilizer
The concentration of P or N in water is an indication of eutrofication by fertilizers. High values indicate poor water quality.

WATER QUANTITY

Outflow to sea
The outflow to the sea is equal to the water available for the ecosystems near the mouth of the river such as mangrove and wetlands. It is also an indicator for the minimum flow during the year and can be used as an industrial indicator for the navigability of the river.

High flow
The occurrence of a runoff that passes a certain threshold for the river. Riverine ecosystems need to be flooded once a year. When this does not happen for some consecutive years, these ecosystems will disappear. The environmental high flow is part of the environmental water resource (EWR).

Low flow
Is the same as outflow to the sea, but more upstream, and ecosystems have a minimum requirement for the amount of water that is available in the EWR.

Decision indicators

% Freedom
Longitudinal and lateral freedom are indicators of the pristinity of the river system. The indicator for lateral freedom provides the percentage of the river area that is not protected by man-made dykes. The higher this percentage, fewer dykes or protective measures have been applied. More dykes generally protect the river from flooding but prevent exchange of nutrients and fish to the flood-plains. This influences biodiversity negatively. Decrease in lateral freedom also influences the flood dynamics of the river basin, and hence the ecosystem equilibrium. Longitudinal freedom relates to the number of dams and barrages in the river. These man-made obstructions hinder fish migration, and hence lower biodiversity and ecological quality. Decrease in longitudinal freedom also influences the flood dynamics of the river basin (high flow, low flow), and hence the ecosystem equilibrium.

No. fish
The number of indicator fish species is a bio-indicator for water quality. The fish need a certain minimal quality of water and quality of habitat (geomorphologic) to be able to survive and reproduce in a river system. If the fish are able to survive, the overall quality of the system and water quality are good. In some rivers, the species can be salmon or trout, but any other fish can be taken. Instead of one fish species, the diversity of fish species can be used.

ha Ecosystem
Forest, wetland, upstream forest, etc. The area of forest species or wetlands in the basin is an indicator for the size of the ecosystem habitat. Large areas have a larger natural value as their capacity to host species (biodiversity) is greater compared with smaller areas. Larger areas also better regulate changes in erosion rates and sediment loads. The area of wetlands is an indicator for the size of this habitat and it measures indirectly the biodiversity of the area.

ha Badland/
desert
The area of badland or desert is an indicator for the size of this non-inhabitable land. An increase in area in the basin means that the lands are deteriorating in quality.

Table 4.1. Environmental adaptation measures in Syr Darya and Rhine Basins.

Measure	Syr Darya	Rhine
Decrease agricultural land	X	X
Use less fertilizer	X	
Use less herbicide	X	X
Change management of dams (minimum flow requirements)	X	
Decrease amount of water for irrigation	X	X
Build fish traps	X	X
Develop sewage treatment plants	X	
Develop dykes	X	
Increase floodplain storage and retention		X
Temperature control devices		X

start 3–4 weeks earlier and will decrease in volume compared to the current situation. This is expected to negatively influence the environment through a decrease in water availability in summer, probably below the EWR of the basin (see Chapter 6 for more information). The set of indicators that best allows monitoring the state of the water resources in the Syr Darya Basin are: 'outflow to the Aral Sea' and concentrations of 'PCB', 'fertilizer' and 'NaCl'. The DIs used are: 'area of badlands/desert' and 'longitudinal freedom'.

In the recent past, the environmental issues in the Rhine Basin mainly relate to pollution of the river, especially during the period between 1960 and 1975. After this period the inflow of pollutants from industry and cities was constrained by the construction of sewage treatment plants and international treaties (Klein *et al.*, 2003). Currently, the most pressing issues related to the environment are: extreme high flows, low flows (which cause an increase in water temperature and an increase in concentration of pollutants), input of pollutants and habitat suitability for the return of fish species like salmon and the return of bird species such as the black stork (de Bruin *et al.*, 1997). For the Rhine the SIs used are: 'concentration of BOD', 'low flow' and 'fertilizer'. The DIs are: 'fish diversity', 'upstream forest', 'floodplain forest' and 'wetlands'.

The change in the state of the water resources in the two basins is simulated for two time slices, 2010–2039 and 2070–2099, and two scenarios, A2 and B2 (see Chapters 2, 5 and 6). First, with the help of models and experts, the values of the SIs and DIs are estimated for the future without considering adaptation. These values are compared with the current situation. A change in DI and SI values indicates an impact. This provides a baseline to which the different adaptation strategies will be compared. Next, adaptation strategies are formulated from various measures, ranging from technical measures to policy measures.

For a water manager in the Syr Darya Basin, the environmental focus would lie on the reduction of pollutants, the supply of more water to the wetlands in the basin and an increase in outflow to the Aral Sea. The resulting set of measures (from here called a strategy) is listed in Table 4.1. For the Rhine Basin, the environmental measures would focus on restoring the habitats in the river and on the floodplains to make them suitable again for supporting certain (fish) species.

In Table 4.1, adaptation measures for the Syr Darya and the Rhine Basins are listed. The measures mainly address the improvement of environmental quality, and are almost all structural engineering measures, as opposed to non-structural measures such as water pricing and efficient water demand management. This could be a result of the time frames and the scope of the ADAPT project, with a greater emphasis on structural than non-structural measures. Structural measures are considered good measures for the long term as opposed to non-structural measures, which are more suitable for the shorter term (Dvorak *et al.*, 1997).

Next, models were used to evaluate the adaptation strategies by calculating new values for the SIs and DIs. The measures and strategies are evaluated by comparing the difference in the DIs for the different adaptation strategies and by comparing them to the pre-set goals. By varying the measures and running the models for these slightly different strategies, the effects will change and the best strategy can be found in an iterative way. For both storylines, A2 and B2, and for the two time slices, 2010–2039 and 2070–2099, the simulation model is run.

For each basin, originally four adaptation strategies were developed, which focused on enhancing environment, food security, industrial production and a strategy that is a mix of these three. In the basin chapters, all these strategies are evaluated through the values of indicators. In Table 4.1, only the values of the indicators for 'no adaptation' (baseline) and the environmental strategy are shown for the two time slices. The effectiveness of the strategies can be evaluated through this table.

From Table 4.2 it can be shown that not all the indicators apply to both basins. For example, the state indicator 'concentration PCB' applies to the Syr Darya, but not to the Rhine. This can be explained by the use of products that contain PCBs. These products have been prohibited for many years in the Rhine Basin, but are still allowed in the Syr Darya Basin. It is also possible for other indicators to be unsuitable because that species or substance does not occur in that region.

So far, only planned adaptation measures have been described. Apart from planned adaptation, spontaneous adaptation will occur. These are the responses of the ecosystems as described earlier. For the environment, the most important spontaneous adaptation is the expected increase in water efficiency by plants, as a result of higher CO_2 concentrations in the atmosphere. In Chapters 5–12, the different strategies and effectiveness of the measures are explained more thoroughly. The migration of species is not taken into account in the development of adaptation strategies and in the evaluation of the strategies, although it is expected to occur in all the basins. With the available data on tree species, for instance, it was not possible to make estimations like the ones made for agricultural crops (see Chapter 3).

Effects of measures as explored above are meant to enhance the quality of the environment, but some of them are not beneficial for the other users in the basin. For instance, increasing the winter flow in the river by releasing water from reservoirs decreases food yield through less irrigation water and reducing the the possibility to produce hydropower in summer. The table clearly shows that a measure that benefits one goal is probably negative for the realization of another, so in the development of adaptation strategies trade-offs have to be made between the different users and their goals.

Table 4.2. Results for the Syr Darya and Rhine Basins for the A2 scenario.

	Rhine				Syr Darya			
	No adaptation		Environmental strategy		No adaptation		Environmental strategy	
Indicators	2039	2099	2039	2099	2039	2099	2039	2099
State indicators								
[BOD]	0	0	+/−	+/−	−−	−−	+	+
Fertilizer	0	0	0	+	−−	−−	+/−	+
[PCB]	n/a	n/a	n/a	n/a	−−	−−	+/−	+/−
[NaCl]	n/a	n/a	n/a	n/a	−	−−	−	−−
Outflow to sea	n/a	n/a	n/a	n/a	n/a	2.1	n/a	3.8
Radioactive pollution	n/a	n/a	n/a	n/a	−	−	+	++
Decision indicators								
No. fish (diversity)	0	0	+	+				
ha upstream forest	0	0	+	++	n/a	n/a	n/a	n/a
ha floodplain forest and wetlands	0/+	+	+	++	n/a	n/a	n/a	n/a
% longitudinal freedom	0	0/+	+	+	25	34	25	30
% lateral freedom	0	0/+	+	+	n/a	n/a	n/a	n/a
ha of badland/desert	n/a	n/a	n/a	n/a	++	−−−	0	−−−

+, Positive effect of the strategy on this indicator (*quantitative effect can be an increase or a decrease!*).
−, Negative effect of the strategy on the indicator.
0, No effect of the strategy on this indicator.

Conclusions

There are many ways through which the environment is influenced by CC, some direct and some indirect. The range of impacts is not certain yet, and only estimations can be made. However, it is certain that ecosystems will be affected; a few studies have already proved that currently changes can be detected in the field. And according to palaeographic studies, in the past large changes have occurred which influenced the environment greatly. Ecosystems can be seen as a water user just like agriculture and humans, and this approach would mean that a certain amount of resources should be allocated for ecosystems. Hence, it is suggested to include ecosystems as an integral part in water management options to ensure they are considered in basin management plans.

The two case studies indicate it is possible to improve the environment by taking measures, even though ADAPT paid relatively little attention to the subject. The AMR approach is suitable for other basins, and it is desirable that analogous to crop modelling the effects of CC on important ecosystem species is simulated. The adaptation strategies that were developed pay little attention to non-structural measures. For a well-balanced strategy these measures have to be taken into account. The scenarios will be more realistic when the non-climatic changes and pressures on the environment are included, because according to the literature the impact of CC on the environment in the coming decades is still much less significant than the impacts from population growth and land use change (Lettenmaier *et al.*, 1999; Morgan *et al.*, 2001; Parmesan and Yohe, 2003). Through an integrated assessment of the pressures on the environment the climatic and non-climatic impacts can be integrated and their combined effect on ecosystems can be assessed.

The pressures on ecosystems are expected to increase because pressures on human society will increase the demand for land, water and wildlife resources. The result is a change in the Earth's land surface, the ecosystem services humans receive, and the landscapes where humans live at regional and global scales (UNEP, 1998). In overcoming these effects good governance is needed, including good management of the WRS. It is shown that the management of the WRS will include a trade-off between the different water users in the basin. The AMR framework is a tool that helps basin managers and inhabitants to support decisions with respect to the different stakeholders involved in the process.

References

Aerts, J.C.J.H., Lasage, R. and Droogers, P. (2003) *A Framework for Evaluating Adaptation Strategies.* IVM research report series, R03-08. Vrije Universiteit Amsterdam, The Netherlands.

Allen, C.D. and Breshears, D. (1998) Drought induced shift of a forest-woodland ecotone: rapid landscape response to climate variation. *Ecology* 95, 14839–14842.

Band, L.E., Scott Mackay, D. and Creed, I.F. (1996) Ecosystem processes at the watershed scale: sensitivity to potential climate change. *Limnology and Oceanography* 41, 928–938.

Barber, V.A., Juday, G.P. and Finney, B.P. (2000) Reduced growth of Alaskan white spruce in the twentieth century from temperature drought stress. *Nature* 405, 668–673.

Blasco, F., Saenger, P. and Janodet, E. (1996) Mangroves as indicators of coastal change. *Catena* 27, 167–178.

de Bruin, D., Hamhuis, D., van Nieuwenhuijze, L., Overmars, W., Sijmons, D. and Vera, F. (1987) *Ooievaar, de toekomst van het rivierengebied.* Stichting Gelderse Milieufederatie, The Netherlands.

Bryant, E. (1997) *Climate Process and Change.* Cambridge University Press, Cambridge.

Caruso, B.S. (2001) Temporal and spatial patterns of extreme low flows and effects on stream ecosystems in Otago, New Zealand. *Journal of Hydrology* 257, 115–133.

Covich, A.P., Fritz, S.C., Lamb, P.J., Marzolf, R.D., Matthews, W.J., Poiani, K.A., Prepas, E.E., Richman, M.B. and Winter, T.C. (1997) Potential effects of climate change on aquatic ecosystems of the great plains of North America. *Hydrological Processes* 11, 993–1021.

Dakova, S., Uzunov, Y. and Mandadjiev, D. (2000) Low flow – the river's ecosystem limiting factor. *Ecological Engineering* 16, 167–174.

Doll, P., Kaspar, F. and Lehner, B. (2003) A global hydrological model for deriving water availability indicators: model tuning and validation. *Journal of Hydrology* 270, 105–134.

Dublin Statement (1992) Dublin Statement. *Environmental Policy and Law* 22, 54–55.

Dvorak, V., Hladny, J. and Kasparak, L. (1997) Climate change hydrology and water resources impact and adaptation for selected river basins in the Czech Republic. *Climatic Change* 36, 93–106.

Ellison, A.M., Mukherjee, B.B. and Karim, A. (2000) Testing patterns of zonation in mangroves: scale dependence and environmental correlates in the Sundarbans of Bangladesh. *Journal of Ecology* 88, 813–824.

Frederick, K.D. and Gleick, P.H. (1999) Water and global climate change: potential impacts on U.S. water resources. Pew Center on Global Climate Change.

Gilbert, A.J. and Janssen, R. (1998) Use of environmental functions to communicate the values of a mangrove ecosystem under different management regimes. *Ecological Economics* 25, 323–346.

IPCC (2001) *Climate Change 2001 – Impacts, Adaptation and Vulnerability. Contribution of Working Group II to the Third Assessment Report of the Intergovernmental Panel on Climate Change.* McCarthy, J.J., Canziani, O.F., Leary, N.A., Dokken, D.J. and White, K.S. (eds) Cambridge University Press, Cambridge.

Kabat, P. and van Schaik, H. (2003) *Climate Changes the Water Rules: How Water Managers Can Cope with Today's Climate Variability and Tomorrow's Climate Change.* Dialogue on Water and Climate, Printfine, Liverpool.

Klein, H., Douben, K.J., van Deursen, W. and de Ruyter van Steveninck, E. (2003) *Water, Climate, Food and Environment in the Rhine Basin.* ADAPT project report, Delft, The Netherlands.

Klein, R.J.T. and MacIver, D.C. (1999) Adaptation to climate variability and change: methodological issues. *Mitigation and Adaptation Strategies for Global Change* 4, 189–198.

Lettenmaier, D.P., Wood, A.W., Palmer, R.N., Wood, E.F. and Stakhiv, E.Z. (1999) Water resources implications of global warming: a U.S. regional perspective. *Climatic Change* 43, 537–579.

Lorenz, C.M. (1999) Indicators for sustainable river management. PhD thesis. Vrije Universiteit, Amsterdam.

McDonald, M.E., Hershey, A.E. and Miller, M.C. (1996) Global warming impacts on lake trout in arctic lakes. *Limnology and Oceanography* 41, 1102–1108.

Mohseni, O., Stefan, H.G. and Eaton, J.G. (2003) Global warming and potential changes in fish habitat in U.S. streams. *Climatic Change* 59, 389–409.

Morgan, M.G., Pitelka, L.F. and Shevliakova, E. (2001) Elicitation of expert judgment of climate change impacts on forest ecosystems. *Climatic Change* 49, 279–307.

Parmesan, C. and Yohe, G. (2003) A globally coherent fingerprint of climate change impacts across natural systems. *Nature* 421, 37–42.

Poff, N.L., Brinson, M.M. and Day, J.W. (2002) Aquatic ecosystems and global climate change: potential impacts on Inland Freshwater and Coastal Wetland Ecosystems in the United States. Pew Centre on Global Climate Change.

Savoskul, O.S., Chevnina, E.V., Perziger, F.I., Vasilena, L.Y., Baburin, V.L., Danshin, A.I., Matyakubov, B. and Murakaev, R.R. (2003) *Water, Climate, Food and Environment in the Syr Darya Basin.* ADAPT project report, Tashkent.

Smakhtin, V.U. (2001) Low flow hydrology: a review. *Journal of Hydrology* 240, 147–186.

Sweeny, B.W., Jackson, J.K., Newbold, J.D. and Funk, D.H. (1992) Climate change and the life histories and biogeography of aquatic insects in Eastern North America. *Global Climate Change and Freshwater Ecosystems*, 143–176.

UNEP (1998) *Handbook on Methods for Climate Impact Assessment and Adaptation Strategies.* In: Feenstra, J., Burton, I., Smith, J. and Tol, R. (eds) United Nations Environment Program, Institute for Environmental Studies, Amsterdam.

Valeo, C., Beaty, K. and Hesslein, R. (2003) Influence of forest fires on climate change studies in the central boreal forest of Canada. *Journal of Hydrology* 280, 91–104.

White, A., Cannel, M.G.R. and Friend, A.D. (1999) Climate change impacts on ecosystems and the terrestrial carbon sink: a new assessment. *Global Environmental Change* 9, 21–30.

Zheng, Z. and Lei, Z.Q. (1999) A 400,000 year record of vegetational and climatic changes form a volcanic basin, Leizhou Peninsula, southern China. *Palaeogeography, Palaeoclimatology, Palaeoecology* 145, 339–362.

5

How Much Water will be Available for Irrigation in the Future? The Syr Darya Basin (Central Asia)

OXANA S. SAVOSKUL,[1] ELENA V. SHEVNINA,[2] FELIX PERZIGER,[3] VIACHESLAV BARBURIN[3] AND ALEXANDER DANSHIN[3]

[1]Institute of Geography Russian Academy of Sciences, Moscow, Russia; [2]Arctic and Antarctic Research Institute, St Petersburg, Russia; [3]Lomonosov Moscow State University, Moscow, Russia

Introduction

The Syr Darya Basin (Fig. 5.1) is one of two major basins in the Aral Sea Basin in Central Asia. It has an area of 402,760 km^2 divided between the four ex-Soviet states of Kyrgyzstan, Uzbekistan, Tajikistan and Kazakhstan. Approximately 20 million people inhabit the basin, 73% living in rural areas, making their living from agriculture. About 55% of the land is used as pastures supporting livestock of sheep, cattle, goats, horses and camels. Furthermore, 8% of the land is used for crop production. The climate in the basin is hot and arid, but in the mountains the climate is more cool and humid. Soils are thin and infertile, but can be productive for certain crops with adequate irrigation, which is not abundant in the region. An immense irrigation network inherited from Soviet times is still in operation, but in part needs renovation and reconstruction in order to increase irrigation efficiency and hence better use of valuable water resources.

After the disintegration of the USSR, the strictly centralized water management system came to an end, and the problems with coordination of water management became a hot issue in the region. There is a distinct conflict of interests between the industrial and agricultural users of water across the different transboundary countries. The major hydropower plants are in the upper stream Kyrgyz Republic, while main irrigated crop land is in the other three countries, thus downstream agriculture is in a more vulnerable position. The largest reservoir in the basin, the Toktogul artificial lake (located in the territory of the Kyrgyz Republic), has the key position because of its location in the upper part of the basin and its capacity to hold more than half of all artificial water storage in the basin. The lake supports a hydropower plant. Before the disintegration of the Soviet Union in 1990, Toktogul lake outflow and hydropower production were regulated taking into consideration the demands for water for irrigation downstream. After

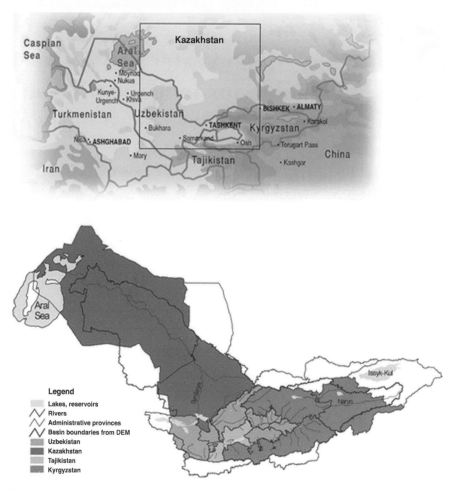

Fig. 5.1. The Syr Darya Basin in Central Asia.

1990, Kyrgyzstan started acting in its own interests, generating more hydropower for domestic needs in the cold period of the year, when the demands are higher, thus drastically reducing water supply to the agricultural areas of Kazakhstan and Uzbekistan in summer (Fig. 2.9). An attempt to settle the crisis was made when Kazakhstan and Uzbekistan both signed bilateral swap agreements with Kyrgyzstan to exchange coal and electricity for water. Though these states failed a number of times to meet the agreed targets, nevertheless, the existing institutions for the regulation of the transboundary water allocation may be considered quite effective in solving the problem.

   The main environmental issue in the Syr Darya Basin remains the collapse of natural ecosystems in the area of the Syr Darya Delta. Here, once productive wetlands have turned into a drying bed of the northern Aral Sea since the 1960s. As a consequence, the fish population of the lake has drastically reduced, virtually eliminating the commercial fishing industry in the region. Furthermore, the exposure of the dried-up bed of the Aral Sea allowed strong winds to erode the underlying sediments, con-

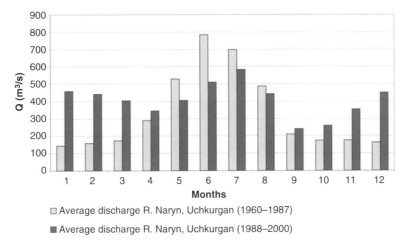

Fig. 5.2. Average monthly Toktogul lake outflow in 1960–1987 (grey columns) and 1988–2000 (black columns).

tributing to a deterioration of air quality for the nearby residents and soil quality due to salt-laden particles falling on arable land. Salinization and waterlogging due to irrigation represent serious threats to irrigated land. The area affected has increased during the last decade from roughly 25% to 50% of irrigated land (Raskin *et al.*, 1992; Heaven *et al.*, 2002). According to Rust *et al.* (2001), presently 31% of the irrigated area has a water table within 2 m of the surface and 28% of the irrigated area suffers from moderate to high salinity levels, which result in a crop yield decline of 20–30%.

The water development system of the region is called 'one of the most complicated water development systems in the world' (Raskin *et al.*, 1992, p. 57). Six large artificial water reservoirs and a number of smaller ones constructed for the purposes of water storage for irrigation and hydropower production have in sum a water storage capacity of 35 km^3. Agricultural water demands by far outweigh those of industry and domestic needs. In the last four decades, the water resources were heavily overexploited, which resulted in a dramatic decrease in the outflow of the Syr Darya to the Aral Sea (see Table 5.1). That, together with overexploitation of water resources in the Amu Darya Basin, led to a dramatic drop in the level of the Aral Sea and an environmental catastrophe in the delta areas of both rivers.

The last decade's changes in the running of the water reservoirs caused not only downstream water shortages over summer, but also an excess of water in winter that

Table 5.1. Syr Darya outflow to Aral Sea.

Period	Runoff (km^3)
Before 1960	50–60
1961–1973	25–30
1974–1987	5–10
After 1988	10–20

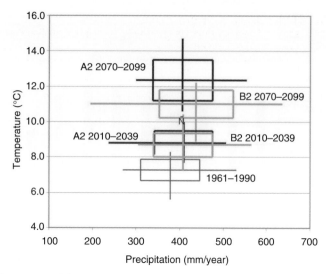

Fig. 5.3. Climate variability: current state and under climate change scenarios. The cross in the centre of each box corresponds to the mean annual values of temperature and precipitation, box length and height equals to two standard deviations, whiskers' outermost points are absolute maximum/minimum of the corresponding variable.

is not used for agriculture. However, the excess water in winter does not reach the Aral Sea. The water flows into an isolated Arnaysay depression, creating a system of lakes totalling 2000 km² in area and raising the groundwater table. As a result, there are widespread newly formed swamps, covering an area of over 20,000 km² in the Arnaysay depression.

Besides these expected changes as a result of internal socio-economic and policy factors, external changes such as climate change (CC) will have an impact on water resources and thus also on the socio-economic situation (UNFCCC, 1999, 2001). We will discuss the possible impacts of CC and assess some adaptations that will allow water managers to cope with these impacts.

Climate Change Projections

Regional climate change scenarios were constructed based on the outputs of the Had3 and ECHAM4 Global Circulation Models (GCMs). For both model outputs, two scenario variants were used: A2 and B2. The time slices considered were 2010–2039 and 2070–2099. Since the original resolution of the GCMs is quite coarse, a downscaling procedure was used to prepare the CC scenarios for regional modelling with a final resolution of 1×1 km² (see Chapter 2).

The analysis of the scenarios (Fig. 5.3) shows a good deal of similarity between outputs from Had3 and ECHAM4 in terms of monthly changes. The Had3-A2 scenario shows the highest temperature increase (5.1°C). The Had3-B2 shows a future with a moderate temperature increase (3.7°C). According to both scenarios, the

Table 5.2. Summary of the mean annual changes of the main climate parameters. dTMP is annual temperature deviation (°C) from the baseline (1961–1990) value; xPRE is annual precipitation increase related to the baseline value (1961–1990).

Time slice	Model	dTMP	xPRE
2010–2039	ECHAM	2.1	1.13
	A2	1.5	1.08
	B2	1.6	1.07
2070–2099	ECHAM	5.4	1.10
	A2	5.1	1.07
	B2	3.7	1.16

summer period is expected to grow more arid, despite an overall increase of annual precipitation in the range of 1.07 to 1.17 times its present value on average. Table 5.2 shows a summary of the projected precipitation and temperature changes.

In the IPCC projections, some key points relevant to Central Asia, and Syr Darya in particular, are:

- for the years 2070–2099, the absolute increase of annual mean temperature will be 4–7°C, while annual precipitation will increase 7–16% as compared to the baseline (1961–1990) interval;
- the temperature variability is expected to increase remarkably: the standard deviation of temperature fluctuations might increase nearly twofold; and
- precipitation variability is expected to significantly increase only under scenario B2 over the period 2070–2099, which also suggests a significant increase of the extremes.

The historic data validate the climate scenarios used in our study: both A2 and B2 scenarios suppose the same tendency for the Central Asian region: overall increase of precipitation as a consequence of global warming. Meteorological observation data in the region (Savoskul *et al.*, 2000) suggest also that variability of climate parameters correlates with climate humidity: the more arid the climate, the less variation in extremes is simulated in the scenarios. For the Syr Darya Basin, it means less climate variability (CV) in the upper reaches of the basin, which is an important note taking into consideration that this is where runoff is formed.

Under the CC scenarios used in this study, the following changes are expected for climate variability. Absolute changes in temperature over the period of 2010–2039 are compared to the ranges of its baseline period variability. For the time slice 2070–2099, the absolute changes of annual means will be far beyond the range of baseline period extremes. At the same time, the temperature variability is expected to increase, e.g. the standard deviation of temperature fluctuations might increase nearly twofold. On the contrary, the absolute increase of precipitation is expected to remain within the range of its current variability. Precipitation variability is not expected to significantly increase. There are no significant changes of standard deviation, apart from scenario B2 in the period 2070–2099, which also suggests a significant increase in the extremes (Fig. 5.3).

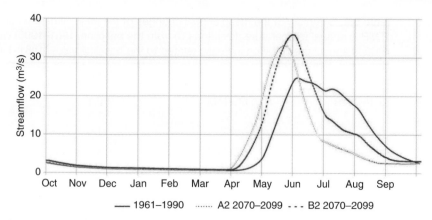

Fig. 5.4. Streamflow (m³/s) modelling for the Charvak sub-basin for the time slice 2070–2099.

Impacts from Climate Change

Hydrology

The impacts of climate change on general basin hydrology have been simulated using the SFM hydrological model (Denisov *et al.*, 2002). Since the changes in temperature and precipitation for the time slice 2010–2039 are virtually the same under both the A2 and B2 scenarios, there is nearly no difference between SFM outputs under A2 and B2 scenarios. For 2010–2039, SFM runs do not give any significant change for the inter-annual runoff distribution. However, there is a pronounced tendency that is much more apparent in the modelled changes of runoff distribution for the time slice 2070–2099, for an earlier onset of spring high waters (shifting it by 5–7 days) compared to the baseline period (1961–1990), sharpening the annual runoff peak in spring and increasing its height, while a slight lowering of streamflow (approximately by 10% as compared to the baseline period) is expected to occur from late June till August. Despite an overall increase of annual precipitation (in the range of 1.07–1.08 of the baseline value) and very insignificant increase of annual runoff (in the range of 1.03–1.04 of the baseline value), on average less water will be available in the period of highest demand for irrigation. However, this is not expected to impose any significant impact on agriculture, since currently existing water management mechanisms in transboundary water allocation allow effective adjustments for a much broader range of year-to-year variations in the availability of water resources (Kipshakaev and Sokolov, 2002).

Over the period 2070–2099, there are remarkable differences in the SFM outputs for the scenarios A2 and B2 (Fig. 5.4). The most drastic changes as compared to the current situation are expected to occur under scenario A2. Onset of high waters in spring is expected to start 3–4 weeks earlier compared to the baseline period (1961–1990). The duration of the annual peak is expected to shorten and the maximum specific runoff is expected to be 25% higher. The changes in the hydrological cycle under the B2 scenario are similar, but less pronounced. The onset of the

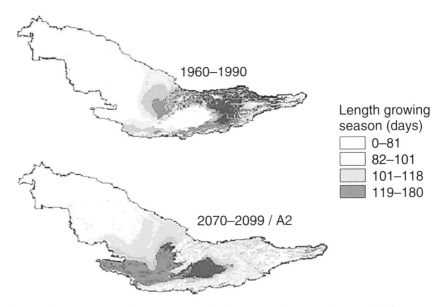

Fig. 5.5. Patterns of LGP for the current situation (1960–1990) and expected changes under the HadCm3-A2 scenarios for the period 2070–2099.

spring high water period would be 2–3 weeks earlier than at present, its peak approximately 30% higher, but the duration of the high water period will decrease less dramatically than under the A2 scenario.

SFM runs show for the A2 scenario a slight reduction (−4%) of annual flow as compared to baseline period, and a slight increase (+7%) for the B2 scenario. The changes of the inter-annual runoff distribution pattern would lead to serious water shortages over the summer period, when the water is mostly needed for irrigation. While at present 68% of the annual flow occurs during 3 summer months (June, July, August), under the A2 scenario this figure would be nearly halved (35%), and under the B2 scenario it would be only 50%. For the time interval 2070–2099, both scenarios impose a very serious negative impact on agriculture, necessitating a very balanced adaptation strategy to mitigate the risks for food production in the Syr Darya Basin. The impacts on industry and environment are also negative since the risk of spring floods might significantly increase, causing danger for dam security and over-flooding in the lower basin.

Environment

Recent patterns of the Length of the Growing Period (LGP) and resulting changes under different climate scenarios are shown in Fig. 5.5 (note different intervals in legend). The numbers in Table 5.3 provide the averages for the basin; more specific conclusions may be drawn based on the analysis of the spatial pattern of modelled LGP changes (Fig. 5.5). Under all scenarios, one can expect positive LGP changes, particularly in the middle reaches in the Fergana Valley and Golodnaya Steppe, where

Table 5.3. Changes in length of growing period (LGP) (days) (top) and percentage of the basin with positive changes of LGP (bottom).

	Period	
SRES scenario	2010–2039	2070–2099
A2	+6	+29
B2	−3	+16
A2	+90%	+66%
B2	+58%	+59%

the main cropland is located. It is especially remarkable under the A2 scenario (dark grey, Fig. 5.5) for the period 2070–2099. However, since agriculture in those areas is highly dependent on irrigation (90% of the cropland is irrigated, while agriculture in total consumes 86% of water resources in the basin), the projected summer water shortages may outweigh the positive effects of LGP changes.

In terms of changes of rangeland productivity, the LGP outputs are less optimistic. Semi-deserts in the downstream part of the Syr Darya Basin (Kazakhstan) and alpine meadows upstream (Kyrgyzstan) are very vulnerable to CC-induced changes. Apart from the LGP, runs under the A2 scenario for the time period 2010–2039, rangelands are expected to suffer overall negative changes for the LGP. Those areas are traditionally used as pastures for semi-nomadic sheep, cattle and horse breeding in Central Asia. Any negative changes in the already low LGP in those rangeland areas would potentially result in a less productive or almost vegetation-free landscape and the reduction of rangeland area in the basin.

Climate variability changes would have mostly negative impacts on food security through increasing the probability of extreme dry years. As shown above, the productivity of ecosystems upstream and downstream may prove rather vulnerable to the extremely dry conditions. The increased risks of spring floods in extremely wet years may have negative impacts on industry and environment due to the risks imposed on dams' safety. The areas in the lower basin, particularly in Ksyl-Orda oblast of Kazakhstan, are potentially prone to the winter–spring flood risks. Of special importance for the environment is the issue of radioactive tailings in Mailu-Suu in Kyrgyzstan. In case of extremely high spring floods, the radioactive pollutants may leak into the streams and contaminate the main flow of the Syr Darya.

Food production

The estimates for the main crop production and yield changes according to FAO data are shown in Table 5.4 (Chapter 12). Those are done assuming a very substantial increase of cropland area and yield increase. However, expert judgement, according to basin team interviews, shows that there is very limited potential in the Syr Darya Basin for increasing cropland area. The assumption reflected in Table 5.4 is that under business as usual (BAU, without application of adaptation strategies), there

Table 5.4. Expected changes of the crop area, production and yield for the main crops in Syr Darya Basin (Chapter 12).

Crop		1998/BAU	2030/BAU	2050/BAU
Cotton	Area (ha)	812,000	882,000	1,030,000
	Production (t)	1,786,400	2,116,800	2,472,000
	Yield (t/ha)	2.2	2.4	2.4
Wheat	Area (ha)	1,079,000	1,373,000	1,473,000
	Production (t)	2,457,900	4,033,000	4,882,300
	Yield (t/ha)	2.3	2.9	3.3
Potato	Area (ha)	84,000	66,000	60,000
	Production (t)	1,106,400	1,828,200	2,117,600
	Yield (t/ha)	13.2	27.7	35.3
Fodder	Area (ha)	1,350,000	1,544,000	1,634,000
	Production (t)	1,350,000	2,161,600	2,614,400
	Yield (t/ha)	1.0	1.4	1.6
Total cropland (ha)		4,088,000	4,978,000	5,327,000
Total cropland irrigated (%)		88	88	88

would not be any substantial increase in the cropland area in the basin (van Dam, 1999).

SWAP modelling for the Syr Darya Basin (Droogers and van Dam, 2004) suggests a significant increase in yields under the A2 CC scenario, due to higher CO_2 concentration levels as compared to the B2 scenario (Fig. 5.6). However, since water availability remains the main factor restricting crop production, we assume that the 'moderate and humid' B2 scenario would be more favourable in terms of yield increase as well as overall crop production compared to the 'hot and dry' A2 scenario. The assumed changes of production for the main crops in the basin according to basin team judgement relate to the reduction of cotton production (the most water-demanding crop in the basin) and switching to less water-dependent wheat, potatoes and fodder crops. This tendency became quite apparent in the last decade, when due to the political changes (the disintegration of the Soviet Union) a new water manage-

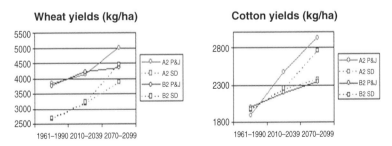

Fig. 5.6. Wheat and cotton yields according to SWAP model outputs (non-dashed line) and Basin team assumptions (dashed line), for CC scenarios A2 and B2.

Table 5.5. Assumed changes of the cropland and rangeland area for the scenarios A2 and B2 in 2070–2099.

	1961–1990	2070–2099 A2	2070–2099 B2
Cropland (ha)	4,141,500	4,100,000	4,500,000
Rangeland (ha)	26,273,800	22,500,000	21,500,000

Table 5.6. Expected changes of the production for the main crops, meat and milk in Syr Darya Basin (basin team expertise). Production is given in ($\times 10^6$ t).

Crop	1961–1990	2010–2039 A2,B2	2070–2099 A2	2070–2099 B2
Cotton	2.240	1.971	2.368	2.756
Wheat	2.660	3.207	3.890	4.470
Potato	1.964	4.422	5.896	6.287
Meat	0.517	0.891	1.180	1.196
Milk	1.676	2.295	2.535	2.727

ment and allocation relationship developed in the newly independent countries of the Syr Darya Basin, which resulted in drastic water shortages for agriculture in the first years. Thus, the adaptation farmers made in the last decade to cope with that change in water availability can be used as a model of response for future changes.

Another assumption is that there would be an increase in livestock and thus an increase in rangeland in the region (Table 5.5). There was an apparent decline in sheep and cattle in the early 1990s, following economic hardship after the disintegration of the Soviet Union. This did not continue, and in the late 1990s the stock numbers gradually increased. Compared to Soviet times, the rangelands in the area were significantly underused in the last decade. Thus, our estimates for the production of meat and milk are indicating positive changes (Table 5.6). Small-scale animal husbandry in subsidiary farms is also a traditional measure for family self-subsistence in the Central Asian region.

Hydropower

Water-related industry in the Syr Darya Basin is confined to hydropower generation only. Since 1970, the Syr Darya (which was navigable up to the mountain foothills) has not been used for transport purposes due to the drastic decrease of river stream flow caused by the overexploitation of water for irrigation purposes and the construction of numerous dams. Population growth in the basin would certainly be related to higher demands for hydropower production. Currently there are several new hydropower plants under construction in upstream Kyrgyzstan territories.

Adaptation Strategies

The development of future adaptation measures in the Syr Darya Basin is to some degree enhanced by analysing the adaptation and adjustment patterns of riparian country economies over recent decades (Sarsembekov, 2000; Dukhovniy, 2001; Tuzova, 2001). Adaptation measures (Table 5.7) can be divided into three categories (E, F, I) according to a water user that is supposed to get most benefit from introducing a measure. E stands for environment, F for food production, I for industry. To estimate the efficiency and relative costs of various adaptation measures (Table 5.7), expert and farmer interviews were used. In order to compare relative costs of the adaptation strategies, we introduced five cost categories. Category 0 was assigned to the measures that involve virtually no cost and are related just to the policies, e.g. open reservoir in winter or change crop pattern. Cost categories 1 and 2 were assigned to relatively inexpensive measures, i.e. introduce water pricing, employ desalinization techniques, increase water productivity, prevent desertification. Categories 3 and 4 were assigned to engineering measures, which are money-, time- and labour-consuming and require major investment, as in the construction of dykes and the reconstruction of irrigation networks. Dam and hydropower plant construction for large water reservoirs was placed under category 4. Estimate of relative costs of an adaptation strategy was made as a sum of the cost categories for each measure, whereas construction of each dam was considered as a separate measure, i.e. relative costs of the construction of three new dams were estimated as 12 (3×4).

An adaptation strategy is defined as a set of adaptation measures. Four adaptation strategies were developed, namely environmental (E), food (F) and industrial (I) strategies, representing the best coping mechanisms for each of three main water users in the basin for minimizing the negative impacts of CC/CV and socio-economic stressors. A mixed adaptation strategy (M) was developed in an attempt to balance the interests of those users. Table 5.8 shows the four adaptation strategies (AS), each comprising different measures. The choice of measures for adaptation strategies was done through expert judgement and fine-tuned with the application of the WEAP model (WEAP, 2002).

Environmental AS

The environmental adaptation strategy includes all the measures designed to mitigate negative impacts of the discussed stressors and to benefit agriculture, i.e. to develop dykes and protection in upper and middle parts of the basin (E1) in order to avert contamination of the Syr Darya by radioactive waste and avoid the risk of flooding in the lower course of the river, to prevent desertification (E2) through the application of sustainable land use practices. Developing sewage treatment plants (E3) will allow improved water quality in the middle and low basin. The low costing measures designed for enhancing food security, i.e. increasing water use efficiency and productivity (F1) and improving water management (F2) are expected to have a positive impact on the environment by increasing outflow to the Aral Sea. Industrial measures to generate hydropower in winter (I2) are closely related to the salinity control and desalinization measure (F6). It would contribute to increasing the outflow to the Aral

Table 5.7. Set of adaptation measures for the proposed CC/CV and socio-economic changes for the Syr Darya Basin.

Measures	Comments	Cost category*
E: Environmental measures (human+nature)		
1. Develop dykes and protection	Is important in Kyrgyzstan and Kazakhstan, i.e. in upper and low basin mostly. Requires capital investment.	3
2. Prevent desertification	Very important measure in arid climates, essential elsewhere, but particularly in upper and lower basin, is based on sustainable land-use policies.	1–2
3. Develop sewage treatment plants	Very important measure for the middle part of the basin.	2–3
F: Food security measures		
1. Increase water use efficiency and productivity	This is actually a set of relatively low-cost measures such as educating farmers, introducing water-saving techniques, searching for the most effective water-use practices at the field scale.	0–1
2. Improve water management	Proven to be an effective measure in transboundary water allocation. The measure also includes water pricing, which at field scale proves effective. In transboundary relationships it depends on political stability and countries' willingness and readiness to implement it.	0–1
3. Change cropping pattern and introduce new crops	Effective and low-cost measure.	0–1
4. Increase water storage capacity and decrease losses in the network	Measure may involve construction of small local water reservoirs, as well as construction of one major Kambarata dam upstream to create a reservoir with storage capacity of 4.6 km^3 mainly for irrigation. The most costly and necessary measure is to reduce leakage in the irrigation network, which involves maintenance and operation costs as well as major investments in network reconstruction.	4–8
5. Increase crop area	There are limited land resources in Syr Darya Basin. Requires a regulation of land rights.	2–3
6. Salinity control/ desalinization	Not very realistic at large scale. Feasible together with I2 measure, since in downstream areas the water used for the generation of hydropower may be utilized for this purpose.	2–3
7. Revive cattle-raising	Very necessary measure: provides security for the local food producers.	1–2

Table 5.7. (continued).

Measures	Comments	Cost category*
I: Industrial measures		
1. Build new reservoirs and hydropower plants	The most costly measure.	4×no. dams
2. Generate hydropower in winter	This measure is the main reason for the conflict of interests between industry and agriculture on the one hand, and upstream and downstream countries on the other. Without a balanced transboundary water allocation policy, may lead to political instability in the region.	0–1

* Cost estimates are divided into five categories: 0, virtually no costs; 1, very low; 2, medium; 3, very costly; 4, extremely costly. For more details, see the relevant text.

Table 5.8. Adaptation measures and strategies proposed for the Syr Darya Basin.

Measures	Adaptation strategies (AS)
E: Environmental measures	**Environmental AS**
1. Develop dykes and protection	E1, E2, E3
2. Prevent desertification	F1, F2, F6
3. Develop sewage treatment plants	I2
F: Food security measures	**Food AS**
1. Increase water use efficiency and productivity	E2
2. Improve water management	F1, F2, F3, F4, F7
3. Change cropping pattern, introduce new crops	
4. Increase water storage capacity, reduce network losses	**Industrial AS**
5. Increase crop area	E1
6. Salinity control/desalinization	F3, F7
7. Revive cattle-raising	I1, I2
I: Industrial measures	**Mixed AS**
1. Build new reservoirs	E1, E2, E3
2. Generate hydropower in winter	F1, F2, F3, F4, F5, F7
	I1

Sea, and is favourable, though not crucial for agriculture, since it diminishes losses in soil productivity but utilizes water that could be otherwise used for irrigation purposes in summer.

Food security AS

The proposed food security adaptation strategy includes the following measures. From the set of environmental measures, preventing desertification (E2) is considered the most beneficial for agriculture, i.e. cattle-raising. The measure is discussed above. The choice of measures designed to directly enhance food production is done with the purpose of meeting agriculture demands for water in two ways: by reducing the actual demands and by increasing the amount of water resources available. Measures F1 (increase water use efficiency and productivity), F3 (change cropping pattern) and F7 (revive cattle-raising) would prove extremely effective under both scenarios by decreasing demands for water. Those measures are favourable for the environment too. To improve water management (F2) is actually a universal measure in terms of coping with water deficits and rationalizing the principles of water allocation between different users. In transboundary relations it may help to make more water available for irrigation. At the local scale this measure may help to save water resources and is closely linked with increasing water use efficiency. The most costly measure (F4) of food adaptation strategy is related with increasing water storage capacity by constructing a major Kambarata reservoir in the headwaters of Syr Darya, and reducing losses in the irrigation network through reconstruction of the irrigation scheme. This requires major investments, but would help to raise more water resources for the demands of crop production. Since the industrial demands for water and the measures to enhance water-related industry in the basin come into conflict with agricultural demands, no industrial measures are included in this adaptation strategy.

Industrial AS

The industrial adaptation strategy is composed of the measures for enhancing hydropower generation, i.e. measure I1 (build new reservoirs) would involve constructing three more new dams and hydropower plants. Together with measure I2 (generating more hydropower in winter), this would enable a significant increase in hydropower production as compared to business as usual. The latter in combination with measure E1 (develop dykes and protection) and less developed crop production would make the industrial adaptation strategy favourable for the environment in terms of improving water quality and decreasing the risk of radioactive contamination. However, in terms of increasing the outflow to the Aral Sea, this strategy is the least effective compared to other adaptation strategies. Food security enhancing measures of this strategy involve only measures of low costs such as change cropping pattern (F3) and revive cattle-raising (F7).

Mixed AS

The mixed adaptation strategy is designed in order to provide a compromise between the interests of three main water users in the basin. Therefore, it includes all the measures favourable for the environment, since those are not in conflict with interests of food production and industry, and a majority of the measures to enhance food production and industrial development, with exception of measure F6 (salinity control and desalinization) and measure I2 (generate more hydropower in winter), which would be in conflict with water demands for agriculture. Measure I1 for this strategy is the construction of two new hydropower plants (A2) or three plants (B2) more than under business as usual.

Assessment of Adaptation Strategies

Indicators

The first step of an assessment of an adaptation strategy is to select criteria or indicators that would allow for quantifying the overall efficiency of an adaptation strategy. The set of indicators proposed for the Syr Darya Basin is given in Table 5.9. The indicators are subdivided into three categories: Environmental, Food and Industrial, each one representing the benefits for a corresponding water resources user. For each indicator the following details are provided: (i) how it is measured, $+++/---$; (ii) what factor (climate change (CC), climate variability (CV), socio-economic change (SE)) is affecting it most; (iii) how it is estimated, e.g. based on a model output or through an expert judgement; and (iv) additional comments.

Business as usual

The second step of the assessment is to select a reference point called business as usual. This is a hypothetical future situation, when no adaptation measures to mitigate negative impacts of CC, CV and SE stressors will be taken. Business as usual is characterized by the status of indicators showing the expected changes in the period 2070–2099 as compared to their status in the baseline period, i.e. 1961–1990. The adaptation strategies will be cross-compared to business as usual.

Under business as usual, the negative impacts of CC/CV and socio-economic stressors as compared to the baseline period can be outlined as follows. For the environment, the major threats are an increase of desert/wasteland, further deterioration of water quality and further drops in the outflow to Lake Aral. For agriculture, an overall increase of crop production is projected because of increasing yields under higher CO_2 levels in atmosphere (Droogers and van Dam, 2004). Also, positive changes in the Length of the Growing Period (LGP) are projected. Although the SWAP model outputs suggest higher crop yields under scenario A2, it is assumed that the B2 scenario is more favourable for the overall crop production, since water availability remains the main constraint in agriculture production (more water availability is projected under B2). Meat and milk production are also expected to increase due

Table 5.9. Proposed indicators for the assessment of adaptation strategies for the Syr Darya Basin.

Indicator	Measured in	Primarily affected by	Estimated	Comment
Environment				
Ha desert/ badland ($\times 10^6$)	+++/−−−	CC	Expert judgement based on LGPM outputs	An important indicator, as rapid desertification is one of the pertinent environmental issues in the lower and upper basin that negatively affects quality and availability of pasture land. Mitigation by adaptation measures rather limited. The negative impacts are given as + (i.e. increase in desert area).
Longitudinal freedom	Number	SE	Expert judgement	
Outflow to Aral Sea	$10^6 \times m^3$	CC, SE	WEAP outputs	Shows whether the ecosystem in Syr Darya delta areas and northern Aral Lake can be kept from further deterioration. Is given as 30-year average value, i.e. represents integrated CC impacts. The higher it is, the better for the environment.
PCB, fertilizer	+++/−−−	SE	Expert judgement	Indicates level of water pollution.
NaCl	+++/−−−	SE	Expert judgement	Indicates level of water pollution.
Food				
Tonnes of cotton	Number	CC, SE	Expert judgement based on LGPM, SWAP and economic growth model outputs	Though it is not food crop, it is currently one of the two important commercial crops in the basin, common for all farm types.
Tonnes of wheat	Number	CC, SE		This is an important commercial crop, common in large-scale farming.
Tonnes of potato	Number	CC, SE		An important crop, for both commercial purposes and for subsistence; very common in small-scale farming.
Tonnes of meat	Number	CC, SE		These two indicators are important for the Syr Darya Basin, since in the upper and lower basin stock-raising is common among the semi-nomadic Kyrgyz and Kazakh people. Meat and milk products
Tonnes of milk	Number	CC, SE		

Table 5.9. (continued).

Indicator	Measured in	Primarily affected by	Estimated	Comment
				play an important role in the traditional diet and are important for subsistence.
Average farm income	US$/year	CC, CE		Indicates level of food security.
No. years with unmet demands for agriculture		CV	WEAP outputs	Unmet demands are determined as less than 75% coverage for agricultural demands, with equal priorities for all other users apart from domestic users.
Industry				
No. of dams	Number	SE	Expert judgement	
Hydropower	+++/− − −	SE, CC, CV	Expert judgement	

to increased rangeland productivity and projected socio-economic changes. As WEAP model outputs demonstrate, the changes in CV, particularly in precipitation, are expected to impose a serious threat to food security. Under the A2 scenario, the risks of unmet water demands in agriculture increase more significantly than under the B2 scenario: WEAP runs indicate 18 years with unmet agriculture demands under A2, and 12 years for B2. Average annual farm income is supposed to be US$2950 under A2 scenario and US$3310 under B2 scenario. For industry, the number of reservoirs and hydropower plants would increase from 25 to 30 and overall hydropower production will increase.

Performance of Adaptation Strategies

Environmental AS

The effects of the environmental AS can be illustrated by the set of indicators and the effects matrix (see Table 5.10): water quality would slightly improve as compared to business as usual, but not entirely because of still high levels of pollution resulting from agriculture. Area of deserts, i.e. rangeland and crop land turned into wasteland, would increase less dramatically than under business as usual. The average outflow to Aral would increase from 4.0×10^6 km^3 to 8.3×10^6 km^3 under the A2 scenario, and from 4.4×10^6 km^3 to 10.1×10^6 km^3 under the B2 scenario. Crop production would increase due to measures F1 and F2, while meat and milk production would be higher due to the measures preventing desertification (E2). Number of years with unmet

Table 5.10. Assessment/impact matrix for the adaptation strategies under CC scenario A2, time slice 2070–2099.

Indicator	Measured in	1961–1990	2070–2099 business as usual	Adaptation strategy 2099 E	2099 F	2099 I	2099 M
Environment							
Ha deserts (×10^6)	+++/---	115,000	+++	++	++	+++	++
Longitudinal freedom	Number	29	34	34	35	37	35
Fertilizer, PCB, NaCl	+++/---	+	++	+	+++	++	++
Radioactive pollution	+ or −	−	+	−	+	−	−
Outflow to Aral Sea	km^3	5.4	4.0	8.3	10.1	7.7	20.9
No. years with flood risk	Number	350	885	−−	+/−	+	+/−
Food							
Tonnes of cotton (×10^6)	Number	2.240	2.370	2.520	2.820	2.430	2.570
Tonnes of wheat (×10^6)	Number	2.660	3.890	3.970	5.100	4.020	4.120
Tonnes of potatoes (×10^6)	Number	1.960	5.900	6.030	7.600	6.370	6.980
Tonnes of meat (×10^6)	Number	0.520	1.180	1.240	1.430	1.300	1.380
Tonnes of milk(×10^6)	Number	1.680	2.540	2.590	3.310	2780	3.030
Average farm income	US$/year	1,500	2,950	3,170	4,060	3,380	3,890
No. years with unmet demand	Number	7	18	9	5	12	6
Industry							
No. of dams	Number	25	30	30	30	33	32
Hydropower	(MW)	60,950	+	++	+	+++	++
Relative costs				8–14	9–15	16–19	24–32

agricultural demands will drop from 18 to 9 under the A2 scenario and from 12 to 6 under B2. Farmers' income is expected to increase insignificantly compared to business as usual. Hydropower production would be somewhat higher than under business as usual due to high production in winter. All the positive effects will be more remarkable under the B2 scenario than under A2. Relative costs of environmental strategy are 8–14, i.e. the least expensive in relation to other adaptation strategies considered below.

Food security AS

The food adaptation strategy is not very effective in terms of satisfying environmental needs. The water quality will remain at the same level as under business as usual, though the outflow to the Aral Sea would increase similarly to that under environmental AS, i.e. to 10.1×10^6 km^3 under the A2 scenario and to 12.2×10^6 km^3 under

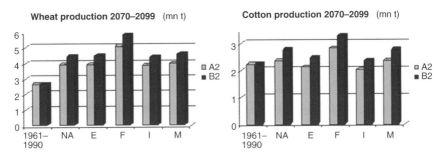

Fig. 5.7. Wheat and cotton production for the four adaptation strategies under CC scenarios A2 and B2. (Baseline period: 1961–1990; NA, business as usual; E, environmental strategy; F, food security strategy; I, industry strategy; and M, mixed strategy.

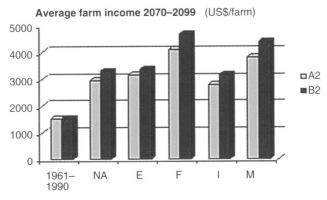

Fig. 5.8. Changes in average farm income under CC scenarios A2 (light grey) and B2 (dark grey).

the B2 scenario (Fig. 5.7). This strategy would make natural ecosystems less vulnerable to desertification. The food adaptation strategy is expected to provide the most essential increase of crop production, particularly of crops less dependent on irrigation, such as wheat and potatoes. Meat and milk production would increase, and expected average farm income is the highest under this strategy compared to other strategies: US$4060 for the A2 scenario and US$4690 for the B2 scenario, which is approximately 1.5 higher than under business as usual (Fig. 5.8). The strategy is apparently the most effective in coping with the risks of not meeting demands for irrigation. WEAP outputs indicate that number of years with unmet agricultural demands would drop from 18 to 5 under the A2 scenario, and from 12 to 3 under the B2 scenario. Hydropower production under this adaptation strategy would remain at the same level as under business as usual. The food adaptation strategy would also be more effective under the B2 climate change scenario, compared to the A2 scenario. The relative costs of Food AS are 9–15, i.e. this strategy requires investment comparable to Environmental AS and nearly half that for Industrial AS and Mixed AS.

Industrial AS

The assessment of the industrial strategy shows a moderate increase of water quality. The expected outflow to the Aral Sea slightly differs from that under business as usual. The industrial adaptation strategy would achieve a meagre amount of 7.7×10^6 km^3 under climate change scenario A2 and 9.7×10^6 km^3 under scenario B2, and there would be no change in the rates of desertification. This strategy would make crop production, particularly wheat and potatoes, and meat and milk production increase, and consequently the average farm income would slightly increase compared to business as usual. Level of food insecurity would remain rather high: the number of years with unmet demands for agriculture would not decrease drastically: it would be 12 instead of 18 years for scenario A2, and 9 instead of 12 years for scenario B2. It should be kept in mind that this effect would be reached not by increasing the amount of water available for agriculture, but by decreasing agricultural demands. Hydropower production would be significantly higher than under business as usual. The costs of the Industrial adaptation strategy are high in comparison with Food and Environmental AS, reaching 16–19 points for scenario A2 and 20–23 for scenario B2.

Mixed AS

Mixed adaptation strategy may be assessed as second to most satisfying in meeting the interests of the three main water users in the basin, i.e. it would help in improving water quality, though less effectively than the Environmental strategy; reducing risks of radioactive contamination; preventing desertification and increasing outflow to the Aral Sea, compared to other strategies to 23.4×10^6 km^3 under climate change scenario A2 and 20.9×10^6 km^3 under scenario B2. The Mixed strategy would provide second to maximum, i.e. to that under Food AS, increases in crop, milk and meat production and in average farm income and raise the level of food security in terms of meeting water demands for agriculture (6 and 4 years of unmet demands for scenarios A2 and B2 correspondingly). Hydropower production would also be higher under this strategy than under business as usual, second to best, i.e. that under Industrial AS. However, the relative costs of this strategy are the highest among all the adaptation strategies under consideration, since it combines most costly measures of all other strategies: thus, relative cost of Mixed strategy would reach 24–32 under scenario A2 and 28–36 under scenario B2. This strategy is also expected to be more effective under the B2 climate change scenario compared with A2.

References

Denisov, Y.M., Agaltseva, N.A. and Pak, A.V. (2000) *Automated Methods of Long-term Forecasting of Mountain Rivers Run-off in Central Asia.* SARHMI, Tashkent (in Russian).

Droogers, P. and van Dam, J.C. (2004) Field scale adaptation strategies to climate change to sustain food security: a modeling approach across seven contrasting basins. IWMI Working Paper. International Water Management Institute, Sri Lanka.

Dukhovniy, V.A. (2001) *Ways of Water Conservation.* SIC-IWCW, Tashkent, Uzbekistan.

Heaven, S., Koloskov, G.B., Lock, A.C. and Tanton, T.W. (2002) Water resources management in the Aral Basin: a river basin management model for the Syr Darya. *Irrigation and Drainage* 51, 109–118.

Kipshakaev, I.K. and Sokolov, V.I. (2002) Water resources of Aral Sea basin – formation, distribution, water use. Abstract of scientific-practical conference, 20–22 February 2002 SIC-IWCW, Tashkent, Uzbekistan. ICWC 10th anniversary, pp. 47–55 (in Russian).

Raskin, P., Hansen, E., Zhu, Z. and Stavisky, D. (1992) Simulation of water supply and demand in the Aral Sea region. *Water International* 17, 55–67.

Rust, H.M., Abdulaev, I., Hassan ul, M. and Horinkova, V. (2001) Water productivity in the Syr Darya River Basin. IWMI internal report.

Sarsembekov, T.T. (2000) Water and food security: a brief review of some world problems of water resources, food production and irrigation. In: Dukhovniy, V.A. (ed.) *Integrated Management of Water Resources.* SIC-IWCW, Tashkent, Uzbekistan.

Savoskul, O.S. (2000) Modelling future environmental change in continental and maritime Asia. Project Report to the Institute of Open Society under the Research Support Scheme.

Tuzova, T.F. (2001) *Water and Sustainable Development in Central Asia, 2001.* Soros-Kyrgyzstan Foundation, Bishkek (in Russian).

UNFCCC (1999) Initial National Communication of the Republic of Uzbekistan under the UN Framework Convention on Climate Change. Phase 1, 1999. UNFCCC, Tashkent.

UNFCCC (2001) Initial National Communication of the Republic of Uzbekistan under the UN Framework Convention on Climate Change. Phase 2, 2001. UNFCCC, Tashkent.

van Dam, J.C. (1999) *Impacts of Climate Change and Climate Variability on Hydrological Regimes.* Cambridge University Press, Cambridge.

WEAP (2002) Water Evaluation and Planning System. Available at: http://www.seib.org/weap

6 Maintaining Sustainable Agriculture under Climate Change: Zayandeh Rud Basin (Iran)

S. Morid, A.R. Massah, M. Agha Alikhani and K. Mohammadi

College of Agriculture, Tarbiat Modarres University, Tehran, Iran

Introduction

Water resources in the Zayandeh Rud Basin in Iran have experienced enormous changes over the last centuries. Water use has been intensified substantially over the last 50 years. The changes are nicely reflected in the name Zayandeh Rud, which means 'the river that feeds itself'. It refers to the times when contributions from groundwater and tributaries increased the river flow in its journey to the outlet point. Nowadays, this picture has changed completely. From each drop of water that is released from the upstream-located reservoir, only half a drop reaches the main city Esfehan, and from this half drop nothing remains at the outlet point of the river.

It is clear that these changes will continue. Population growth is still on the rise, putting more emphasis on domestic water requirements. Industrial activities are becoming more profound, along with their associated water requirements. Food demands from the increased population will also put a higher claim on water, and agriculture (already the biggest consumer of water) might come under severe stress.

On top of these stressors are the uncertain impacts of climate change and climate variability and the need for adaptation strategies. Unfortunately, climate change issues have not been sufficiently considered by the basin's policy makers. At the national level a few activities have been initiated by the National Environmental Organization. We hope the present research work helps in developing a more structured policy with regards to the impacts of climate change and variability.

This chapter first illustrates the characteristics of the basin. Next we will describe some novel approaches, in the context of Iran. A framework is used to assess the impact of climate change and to explore adaptation strategies to minimize the possible negative effects of climate change. We will demonstrate our approach for the Zayandeh Rud Basin in the centre of Iran. Since agriculture plays such an important

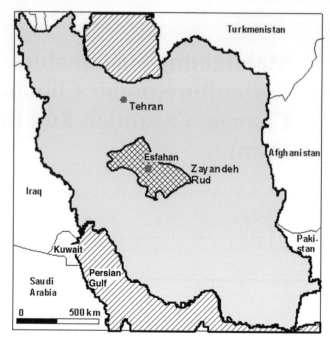

Fig. 6.1. Location of Zayandeh Rud Basin in Iran.

role in the basin, we will concentrate on this, but will do this in the context of other water consumers.

Physical characteristics

The Zayandeh Rud Basin is located in the central part of Iran, with geographical coordinates of approximately 33° North and 53° East (Fig. 6.1). The area of the basin is about 42,000 km². Esfahan province constitutes 88% of the Zayandeh Rud Basin. The rest of the basin is located in Bakhtiyary and Yazd provinces. The city of Esfahan is the capital of Esfahan province, which is one of the oldest cities in the world with about 2 million inhabitants.

The river itself has provided the basis for important economic activities for centuries, including the growth and establishment of Esfahan as the former capital city of Persia. The region has been able to support a long tradition of irrigated agriculture in order to meet the domestic needs of the population and industrial demands.

However, the agriculture sector, being the main water consumer using more than 80% of the available water resources, is heavily under pressure. Numerous factors, including the continued growth of the urban population, the development of new agriculture lands and the rapid increase in industrial demands, have caused water shortages over the last half a century. To overcome these problems, a number of trans-basin water projects were realized and exploited over the last decades. There is, however, still insufficient water to irrigate the total irrigable area. It has also resulted in a reduction of water quality of the Zayandeh Rud, especially downstream of the city of Esfahan. Deterioration of water quality causes problems for the ecosystem of the river and for Gaw Khuny swamp, the final outflow point of the river.

The Zayandeh Rud Basin consists of seven sub-basins, namely Plasjan, Shur Dehghan, Khoshk Rud, Morghab, Zar Cheshmeh, Rahimi and Gaw Khuny swamp. The upper catchment of the basin is part of the Zagros Mountains, with high altitudes and ample precipitation. The general slope direction of the upper basin is west to east and the elevation varies between 1000 m and 3600 m. The upper basin is of paramount importance in terms of water resources, since almost all water that is used in the basin originates from this mountainous area. The natural vegetation covering the lower basin is sparse, as precipitation downstream is very low (50–200 mm/year) and erratic.

Elevation has a significant effect on climatological conditions and its spatial and temporal variation. Precipitation in the basin is mainly driven by the Mediterranean rainfall system, which enters from the north-west of the country. The western mountains of the basin and their direction induce substantial rainfall. Rainfall reduces rapidly in the basin from west to east, ranging from 1400 mm on the most upper portion of the catchment (mostly as snow) to 700 mm in the intermediate part. The amount of rainfall reduces to less than 100 mm at the downstream Gaw Khuny swamp. Potential evapotranspiration ranges from 1450 mm to 2800 mm.

The dominant land use types in the basin are pasture and uncultivated land. The Morghab sub-basin is very important for agriculture and the main irrigation networks are located here. Most of the forests and rain-fed areas are located in the Plasjan sub-basin. So, in terms of economic activities these two sub-basins are the most crucial ones in the Zayandeh Rud Basin.

Water resources

Water resources are highly regulated by man-made infrastructural systems. Numerous water projects have been constructed, are under construction or are under study. The Chadegan Dam is the main water reservoir with a capacity of 1450×10^6 m^3 and has been exploited since 1971. After the construction of the dam, 90,000 ha were added to the traditional network. At present about 297,000 ha of irrigated land is surface water and groundwater dependent.

Even after the construction of the dam, insufficient water is available and water scarcity is common in the basin. Inter-basin transfer is a common practice to alleviate the severe water shortage. For example, water is diverted from the Karoon Basin to Zayandeh Rud by two tunnels. These tunnels divert $300–400 \times 10^6$ m^3 water/year. In spite of this huge project, the basin is still under threat and two additional tunnels are under construction. A third tunnel will divert water from Karoon River and Tunnel No. 2, the Lenjan tunnel, from Dez River upper catchment (Anonymous, 1993). These two new tunnels, with a total capacity of 425×10^6 m^3, are expected to be completed in 2004 and 2008, respectively. Another tunnel, the 75 km Behesh Abad is under study, but investment costs will be huge. The total diversion of water from this tunnel is estimated to be $700–1000 \times 10^6$ m^3. It would join the Zayandeh Rud River downstream of the Chadegan Dam. These projects will play a crucial role in minimizing water deficiency in the basin and the possible negative impacts of climate changes.

Alongside these efforts to alleviate water scarcity by transferring water from outside the basin to the Zayandeh Rud, there are projects diverting water from the basin to neighbouring cities. The first phase of the Yazd project has been operational since 2002, diverting 42×10^6 m^3 water/year out of the basin. In the second phase of the project the amount of diverted water will increase to 78×10^6 m^3/year. Kashan and Shahr Kurd diversion projects with a total capacity of 24×10^6 m^3/year are under construction.

Groundwater resources play a crucial role in storage and regulation of water resources. Wells and qanats, the traditional Iranian system to extract groundwater, are the main means to extract groundwater. Presently, about 20,138 deep and semi-deep wells (total 3224×10^6 m^3/year), 1726 qanats (313×10^6 m^3/year) and 1613 springs (82×10^6 m^3/year) are extracting a total of 3619×10^6 m^3 groundwater/year.

Comparing surface water extraction (1245×10^6 m^3) with that of groundwater (3619×10^6 m^3) reveals the significant role of groundwater resources for the study area. Overall, about 90% of the total water resources in the study area are controlled (Anonymous, 1993).

The Esfahan Water Authority (EWA) studies the basin groundwater budget every year. The year 2000 study revealed that all the plains have a high rate of overdraft. One of the main reasons is that during 2000 the basin suffered from a severe drought. But also the long-term trend demonstrates an overdraft and declining groundwater tables. Return flows from the irrigation systems are the main source for groundwater recharge.

The Zayandeh Rud River has for centuries provided the basis for important economic activities. Looking at water consumption, these activities can be categorized into three sectors: agriculture, industry and domestic. Agriculture is the dominant water user, consuming more than 80% of the river yield. But this amount is still insufficient to irrigate the total irrigable area. It is estimated that water consumption per hectare varies from 10,000 to 14,000 m^3.

The total area of irrigation systems is estimated to be about 180,000 ha, with Neku Abad, Abshar, Borkhar and Rudasht the major irrigation systems in the basin. The Neku Abad and Abshar irrigation systems were constructed in 1970. The designed command area of these systems is about 90,000 ha. Borkhar and Rudash, with a total command area of 83,000 ha, have been under cultivation since 1997 and parts of the systems are still under construction. There are irrigation systems that are expected to be exploited in the near future (e.g. Keron irrigation system).

Huge industrial complexes are located in the basin. The most important ones are Esfahan Steel Mill, Mobarekeh Steel Complex and a number of textile factories that consume about 100×10^6 km^3 water/year. The population of Esfahan city is about 2 million, with a domestic water use of about 80 m^3/year per capita. In terms of servicing the ecosystem, 70×10^6 km^3 is the minimum water requirement for preserving the downstream located Gaw Khuny swamp (see also the next section).

Agriculture and environment

The dominant crops in the basin are: cereal (wheat, barley, rice, maize and sorghum), forage (lucerne, clover, sainfoin and maize), pulse (bean, lentil and chickpea), and

Table 6.1. Main agricultural crops in the basin and average and maximum obtainable yields.

Crop type	Area (ha)	Average yield (kg/ha)	Maximum yield (kg/ha)
Wheat	78,995	4,547	9,000
Barley	28,763	4,418	7,000
Rice	7,698	4,828	10,000
Potatoes	21,807	26,256	50,000

industrial crops (cotton, sugarbeet, safflower and potatoes). Wheat, rice, barley and potatoes are the main staple crops in the basin and a major source of caloric intake for people and livestock (Table 6.1). The yields provided in the table for these staple crops are average values and some farms in the basin have higher performances.

The Zayandeh Rud River and the Gaw Khuny swamp are two important natural ecosystems in the basin. The Gaw Khuni swamp is the final outflow point of the river and is an important wetland recognized by the Convention of Ramsar (1975). The mean area of the swamp is about 43,000 ha, but varies annually as a result of variation in total inflows. During wet years, the depth of the swamp reaches 1.5 m, but normally the depth is between 0.3 and 0.6 m. Quantity and quality of the river are highly dependent on releases from the dam and on the numerous water diversions along the 350 km reach of the river to the swamp. Furthermore, the water quality of the river and swamp is affected by the return flows from the upstream demand sites. The wildlife of the swamp depends on the water depth. The lowest critical depth for vital activity is about 15 cm, which can be maintained with 2.2 m^3/s inflow to the swamp (~75×10^6 km^3 water/year). A more favourable depth is 30 cm, which requires an inflow of about 4.5 m^3/s (140×10^6 km^3 water/year). This depth is optimal to support fish, birds, plants and small mammals (Moeinian, 2000).

Main sources of river pollution can be categorized into three groups: domestic effluents, industrial and agricultural return flows (Anonymous, 1993). Eighteen stations observe the limnological parameters of the Zayandeh Rud river and are maintained by the Esfahan Water Authority and Esfahan Environment Organization. The river water quality can be categorized in four segments. The first segment is from the Chadegan dam up to the upstream of Esfahan (180 km). For this part, main sources of pollutants are agricultural return flows. In the second part, significant changes can be seen on water quality (180 to 220 km). Return flows from industry, especially textile factories, and Esfahan water treatment plants are the main sources of pollutants. The third segment is from 220 to 270 km, where some decay of pollutants takes place. After this segment, agricultural return flows are again the main sources of pollutants that deteriorate water quality, where the Segzy drain is one of the main culprits.

The largest source of domestic pollutant of the Zayandeh Rud River is the effluent of Esfahan city. The wastewater treatment plant of the southern part of Esfahan is in the vicinity of the river. This system, which was designed for 800,000 people, consists of three separate units. The total volume of effluents is some 126,400 m^3/day, causing serious environmental problems in the river ecosystem, especially during low flows. Although water treatment purification of the system reduces the organic discharge of the wastewater, purification efficiency is only about 85% and needs improvement. Downstream of Esfahan city, domestic pollutants are less relevant and river water quality is more affected by agricultural drainage.

Agricultural drainages contain soluble salts, insecticides and herbicides residues, leached chemical fertilizers and heavy metals. The Zayandeh Rud Basin has three main agricultural drainage systems, referred to as Steel Mill, Rudasht and Segzi drains. These drains enter the river at distances from the regulating dam of 111, 254 and 296 km, respectively. The Steel Mill drain is an open channel to control the groundwater level of the Zarrin Shahr region located at the western part of Esfahan city. It receives high amounts of return flows from agricultural lands and conveys them to the river with an approximate flow rate of 14,400 m^3/day. It also receives part of the outflow from the Esfahan Steel Mill water treatment plant. The Rudasht drain collects drainage water of the Rudasht region and discharges it in the river at 5808 m^3/day. The Segzi drain is the largest and most important drain of the basin and the discharge of the drain exceeds the river flow during some periods, particularly in summer and autumn. Mean drain discharge is about 28,700 m^3/day and salinity levels are too high to be used for irrigation.

Socio-economic and institutional characteristics

From 1966 to 1991, the population in the basin increased from 1.1 to 3.0 million. According to the 1996 census, the population of the entire basin was 3.9 million, of which 2.9 million people live in the urban areas (34 cities) and 1 million are rural residents (1212 villages). Most of the residential areas and almost 82% of the basin's population (cities as well as villages) are located along the Zayandeh Rud River. About 76% of the rural population lives in the Morghab sub-basin. The population of Esfahan and its suburbs is about 2 million and their drinking water is directly supplied by the Zayandeh Rud River. The government has taken positive measures to control the population and its growth rate has started to decrease since 1991 and is estimated to be about 2% for 2000.

The main institution responsible for domestic water exploitation and distribution is Esfahan Water Authority (EWA), which is supervised by the Ministry of Energy. The water distribution up to the tertiary irrigation channel level of the irrigation system is also the responsibility of the Ministry of Energy. Supervision of the exploitation from groundwater resources is included in their mandate as well.

The Esfahan Agriculture Authority, which is supervised by the Jehad-Keshavarzy (Agriculture) Ministry, coordinates the water distribution in tertiary and lower level channel networks. Watershed management and small-scale water projects (e.g. groundwater artificial recharge) are some of the related water duties of the Esfahan Agriculture Authority. Environmental issues in the basin are controlled by the Esfahan Environment Authority. The Iranian Environment Organization is an independent organization, which is directly under the supervision of the Iranian President.

Non-governmental organizations (NGOs) have been activated recently in Esfahan, focusing on Zayandeh Rud River and the river ecosystem. In addition to this, several NGOs have invested in minor irrigation systems and maintenance of the systems.

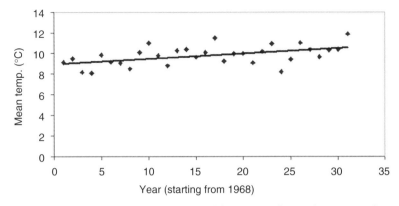

Fig. 6.2. Mean annual temperatures and trend for Damaneh Freydan meteorological station.

Climate Change Projections

Meteorological conditions, such as temperatures and rainfall, have a direct impact on water resources in the basin. Looking at trends in the past can be useful as a first indicator of what possible consequences climate change might have on water resources and, consequently, on agriculture, environment and domestic and industrial water use (Morrison *et al.*, 2000).

As was pointed out earlier, most of the water resources in the basin originate from upper sub-basins (mainly Plasjan sub-basin). Records from the Damaneh Freidan station are a good indicator for the climatic conditions and for climate change in these areas because of its high altitude (2340 m). Figure 6.2 shows mean annual temperatures with a linear trend line analysis. For this station and other stations in the vicinity, the mean annual temperature increase is 0.03–0.05°C/year, although for a few stations in the basin this upward trend was not observed.

From the seven available GCM projections in the IPCC (Intergovernmental Panel on Climate Changes) data set, the Hadley GCM has been selected to be used for further analysis. The A2 and B2 projections based on the most recent IPCC emission scenarios, the so-called SRES (Special Report on Emission Scenario), will be used for two time slices 2010–2039 and 2070–2099.

The Hadley GCM projections are based on grid cells of $2.5° \times 3.75°$ and were downscaled to $0.5° \times 0.5°$ grid using method 2 described in Chapter 2. Subsequently, these projections were transferred in such a way that the main statistical properties of historical measured data match the GCM outputs. For this, the 1972–1990 observed and modelled temperature and precipitation data were used to derive adjustment factors that were subsequently applied to the future projections (2010–2039 and 2070–2099).

Results of applying the above methodology reveal that according to the Hadley GCM projections, the Zayandeh Rud Basin will not experience significant changes in meteorological variables for the first period 2010–2039. However, for the second period 2070–2099, the basin will face more drastic changes. According to the A2

Table 6.2. Statistical parameters of rainfall, mean temperature and discharge according to climate change scenarios.

Period	Statistical parameter	Rain (mm)	Temperature (°C)	Q (m³/s)	Rain (mm)	Temperature (°C)	Q (m³/s)
1971–2001	AV	1458	10.0	45			
	SD	371	0.9	14			
	MAX	894	24.8	289			
	MIN	0	−7.7	6			
Scenarios		A2	A2	A2	B2	B2	B2
2010–2039	AV	1470	11.0	44	1427	11.1	45
	SD	538	0.9	31	361	0.6	34
	MAX	809	27.5	166	632	27.1	170
	MIN	0	−6.4	16	0	−3.5	1
2070–2099	AV	1224	14.6	43	1309	13.2	43
	SD	377	1.1	32	442	0.6	40
	MAX	776	31.2	220	521	30.4	253
	MIN	0	−5.0	10	0	−2.5	6

scenario, mean annual temperature is expected to increase by about 4.5°C and mean annual rainfall depth to decrease by 234 mm. Under the B2 scenario, the temperature increase is 3.2°C and rainfall decrease is 149 mm over the total period 2070–2099 (Table 6.2).

Impacts from Climate Change

Modelling framework

Three different types of simulation models are identified to be necessary for a proper assessment and exploration of adaptation strategies: rainfall–runoff simulation, water allocation programming and crop production prediction. Artificial neural networks (ANNs), ZWAM and SWAP (van Dam *et al.*, 1997) have been selected as the appropriate models to be used.

Rainfall–runoff modelling is required to analyse the processes leading from rainfall to runoff that can eventually be used for irrigation purposes. Several approaches exist, ranging from more physically based methods using semi-distributed hydrological models to simplified rainfall–runoff statistical regression models. We have selected here to use an artificial neural network approach, since this was already developed, well-tested and validated for similar simulations (Dawson and Wilby, 1998; Sanjikumar and Thandaveswara, 1999; ASCE, 2000).

An ANN can be described as an information processing system that is composed of many non-linear and densely interconnected processing elements or neurones. ANNs have the ability to extract patterns in phenomena and overcome difficulties due to the selection of a model form such as linear, power or polynomial.

An ANN algorithm is capable of modelling the rainfall–runoff relation due to its ability to generalize patterns in noisy and ambiguous input data and to synthesize a complex model without prior knowledge (Dawson and Wilby, 1998; Coulibaly *et al.*, 2000).

For rainfall–runoff simulation, 31 years (1972–2002) of records from the Chadegan Dam gauging station that measures total inputs to the dam (upper catchments and transferred water from Tunnel No. 1 and Tunnel No. 2) have been used to train and test the model. Applying different inputs, ANNs models and architectures, the recursive Elman Networks with 7–2–1 architecture was found to be suitable for the study area.

Water allocation between and within different sectors is of paramount importance in Zayandeh Rud. The basin is highly developed in terms of water resources and any change in water allocation has a direct impact on other water users. To deal with these issues, the Zayandeh Rud Water Allocation Model (ZWAM) was developed for this study. The model is able to simulate different water allocation policies, dam operations, environmental issues and examine different scenarios for future changes in the study area. The model is node oriented and the main water demand sites along the river have been embedded in the model. ZWAM is based on similar concepts to the WEAP model (WEAP, 2002).

The agro-hydrological analyses at field scale have been done using the Soil–Water–Atmosphere–Plant (SWAP) model. The model is a physically based one for simulating water, heat and solute transfer in the saturated and unsaturated zone. The model is also capable of simulating crop growth using meteorological data, irrigation planning and phonological crop data. A more detailed description of the model can be found in van Dam *et al.* (1997).

Indicators

To describe the current state of water resources in the basin as well as its future status, a number of indicators have been selected that describe the state of the environment (mainly wetlands) and food security (Aerts *et al.*, 2003).

The indicators that can quantify the state of food security are water allocated to agriculture (m^3/year) food production (tonnes/year) and crop-derived energy production (calories/year). The last indicator makes it possible to compare food production from different crops. Furthermore, since environmental quality is mainly a function of water availability and amount of water that reaches the swamp, it has been decided to use three environmental indicators as 'years with inflows $< 75 \times 10^6 \, m^3$', 'years with inflows $> 75 \times 10^6 \, m^3$ and $< 140 \times 10^6 \, m^3$' and 'years with inflows $> 140 \times 10^6 \, m^3$'.

In summary, the following indicators are used to express the current state and the expected state in the future with and without adaptation strategies:

- water allocated for food production (m^3/year);
- total food production (tonnes/ year); and
- total food energy production (kcal/year).

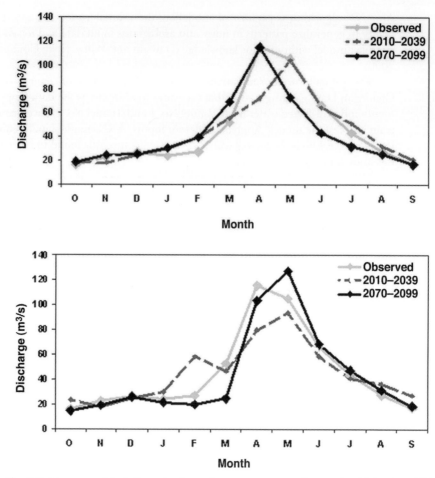

Fig. 6.3. Mean monthly inflows in the main reservoir according to the A2 (top) and B2 (bottom) scenarios.

Impacts of low-flow years on food and environment

The impact of climate change on water, food, industry and environment has been assessed by using the modelling framework as described above. Note that impacts alone without explicit adaptation can be considered the 'business as usual' adaptation strategy.

Besides the impact of climate change also other expected changes, such as population growth, increasing domestic water demands, increasing food requirements and growth in industrial water demands, have been included. A few other drivers are not included, of which the most important one is technological innovation, including the introduction of crop species that are more resistant to water shortage, water salinity or crops that are high-yielding.

Overall, the analysis shows that the basin is under threat of climate change (Table 6.2). The change in average basin rainfall and the temperature increases lead to a

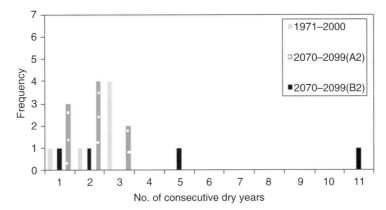

Fig. 6.4. Frequency of number of successive dry years for historical and climate change periods.

reduction in water resources quantity and quality. With respect to the future population growth, the basin's domestic water requirement will reach 344 and 540×10^6 m^3 in 2039 and 2099, respectively (about 150×10^6 m^3 at present). The growing rate of industrial water demand has been assumed to be 1% up to 2010, resulting in about 115×10^6 m^3 by 2010. It is assumed to remain constant after 2010. This assumption is based on the water conservative policies for industry.

Using the trained ANNs model, the streamflows have been simulated for the selected 2010–2039 and 2070–2099 periods for both the A2 and B2 scenarios. The mean monthly flows and their distribution under the A2 scenario show significant changes in timing and volume compared to the historical flows (Fig. 6.3). But, the changes under the B2 scenario are lower and stream flows show almost the same temporal pattern. The sequences of successive dry and wet years have been estimated and are shown in Fig. 6.4. From this figure it is clear that the maximum number of successive dry years during the observed period is 2 years, and increases to 11 years for the A2 scenario and to 3 years for the B2 scenario.

Considering the changes in rainfall and river flows under the climate change scenarios, estimates were made for the groundwater budget for the periods 2010–2039 and 2070–2099. Results show that for most sub-basins an overdraft can be expected. For instance, Najaf Abad aquifer may experience a 2.6 m/year lowering of the water table. The analysis indicates clearly that an increase in exploitation of groundwater is not possible at all, unless new sources (e.g. transfer of water from adjacent basins) for recharge are to be implemented. In general, scenario B2 for the period 2010–2039 shows a declining trend in groundwater level, and this was even more profound for the A2 scenario for the period 2070–2099.

Climate change impacts on agricultural production are the result of two processes, increasing air temperature (i.e. increasing transpiration) and CO_2 enrichment of the atmosphere. These two have been separately considered in this study. In the first step, the SWAP model has been run for the staple crops including wheat, barley, rice and potatoes, and the current climate situation as well as the selected future periods, with scenarios A2 and B2. The general pattern is that considering solely the variation in temperature and rainfall, there is not much change in relative yield for the period 2010–2039 compared with the current situation. However, for the period 2070–2099, 10–15% reduction in crop yields can be expected.

Contribution of CO_2 on crop yields was considered next. It is expected that CO_2 levels in 2100 will reach 640 ppm, which means an increase of more than 300 ppm compared to the present situation (Parry *et al.*, 1999; IPCC, 2000). Responses of plants to this phenomenon will be different. It is expected that C3 plants, such as wheat, barley, rice and potatoes, will respond more positively to rising levels of CO_2 (see also Chapter 3). In a case study in Tabriz (north-west Iran), Koochaki and Mahallati (2001) also reported an increase in crop yields due to CO_2 enrichment. Based on the above studies, it has been assumed that the combined changes of rainfall and CO_2 enrichment will increase crop yields up to 25% in the study area. However, this is based on field-scale analysis and it should be emphasized that this is only potential yield. In addition to potential yield, the actual yield is a function of irrigation water quality and quantity that are anticipated to decline on a basin-level scale in the future.

Evaluation of Impacts and Adaptation

Business as usual

The impact of climate change and associated other drivers without adaptation can be considered as the 'business as usual' adaptation strategy. Using the aforementioned drivers and indicators, the simulation models were applied to calculate indicator values. The status of the basin will be evaluated separately for the selected periods and climate change scenarios.

According to the present water allocation policies, domestic and industrial demands have first and second priority, respectively. The agricultural and environmental sectors are next, such that during the recent prolonged drought spells (1997–2001) water was completely cut for the Gaw Khuni swamp. According to new regulations, Esfahan Water Authorities has been committed to allocating $75–140 \times 10^6 \, m^3/year$ for the river and swamp ecosystems, depending on the year's wetness. These policies have been embedded in ZWAM to examine possible water deficits and their durations during future periods. So, whatever happens with the climate, the domestic and industrial water demands must be fulfilled first and what is left over can be used for agriculture and environment.

The agriculture sector is the main water consumer. While this sector absorbs 80% of the basin water resources in the baseline period (1960–1990), this has to go down in the future since urban and industry have higher priorities. Model results indicate that these agricultural extractions will go down to 75% and 68% of the current situation for the near future (2010–2039) and distant future (2070–2099), respectively. These numbers assume fixed agricultural demands and no new source of water from trans-basin transfer.

The reduction in water availability makes it essential to apply adaptation strategies. Three different possible adaptation measures have been analysed in this study:

- food;
- domestic and industrial; and
- balanced between food and domestic/industrial.

Food-focused adaptation

There is a clear need to produce more food in the future to feed the increased population. Two sub-types of adaptation measures have been evaluated to explore options for agriculture:

- change in the total cropped area in the basin; and
- change in the cropping pattern.

Available data on crop yields are not irrigation-system dependent, but are aggregated at the level of the regions that have been defined by the Esfahan Agriculture Organization (EAO). We have elected to concentrate the analysis on the Nekou Abad and Abshar irrigation systems that are located in the Esfahan region. Rice and potatoes have been analysed according to the Nekou Abad situation and wheat and barley according to that of Abshar. Cropped area for rice, potatoes, wheat and barley are estimated to be 11,260, 3480, 20,892 and 4273 ha, respectively. Optimum water supplies to these crops are estimated to be 17,000, 11,000, 9000 and 8000 m^3/ha, respectively.

To perform our analyses on food adaptation strategies, the basin level and field level models were linked to indicate water quality and quantity at the irrigation systems and the response of crops to the allocated water for the climate change scenarios. During the 2070–2099 period, a negative impact on crop production (average as well as variation) can be expected and also that rice is more sensitive to the climate change than the other crops in the basin.

In order to compare adaptation strategies we will concentrate on water consumed by the crops and the caloric production as indicators. This caloric production was calculated assuming that 3600, 760, 4000 and 4000 are the kilocalories produced per kilogram of rice, potatoes, wheat and barley, respectively.

One of the possible food adaptation strategies is to reduce the total cropped area in order to maintain sufficient irrigation water. This strategy has been investigated and cropped areas have been reduced such that optimum water requirements can be met for the major crops. The results of this strategy are shown in Table 6.3, and indicate that reducing the cropped area in order to have sufficient water per hectare to irrigate crops with the optimum amount has only a positive impact on rice. For the other crops, providing somewhat less irrigation but keeping the area constant is more beneficial. This can be explained by the fact that deficit irrigation is damaging for rice, but is less harmful for the other crops.

Changing the cropping pattern is one of the other strategies that has been explored. Rice requires a substantial amount of water and caloric production is not as high as that for wheat. Moreover, the coefficient of variation of rice yields is higher than that of wheat: 0.14 to 0.29 for rice versus 0.59 to 1.21 for wheat. So, also from a food security point of view, rice production is a less reliable food source. Therefore, the models have been set up to explore what will happen if rice is replaced by wheat.

Table 6.3. Produced total energy (in kcal×10⁹/year) for the two food scenarios: fixed area (=less irrigation per ha) and optimal irrigation (=reduce area).

		A2				B2			
		2010–2039		2070–2099		2010–2039		2070–2099	
	Reference 1990–2000	Fixed area	Optimal irrigation	Fixed area	Optimal irrigation	Fixed area	Optimal irrigation	Fixed area	Optimal irrigation
Rice	196	121	133	53	56	113	114	56	79
Potatoes	69	91	85	66	62	78	71	74	67
Wheat	385	407	367	306	225	380	344	322	242
Barley	76	81	74	60	54	76	71	63	58
Total	726	700	658	485	397	647	600	515	446

Table 6.4. Produced total energy (in kcal×10⁹/year) by changing crop pattern and substitution of rice by wheat.

	A2		B2	
	2010–2039	2070–2099	2010–2039	2070–2099
Wheat (substituted)	390	287	330	309
Potatoes	91	66	78	74
Wheat	407	306	380	322
Barley	81	60	76	63
Total	969	719	864	768

Table 6.4 shows that this is indeed an adaptation strategy that will increase the total amount of food, expressed in calories.

However, rice is more profitable for farmers in terms of revenues in comparison to the other crops. The domestic price of rice in the basin is much higher than current world market prices. If the government wants to minimize rice production in the basin they will have to take measures to make growing crops like wheat more appealing. Such a strategy can result in an increase of 33–48% in the total produced calories in the basin and can reduce agricultural water demands by up to 10%. Since it is not expected that consumers are willing to replace rice for wheat, rice should be imported to the region and the additional amount of wheat produced should be exported. A full economic analysis is required to evaluate the impact of such a measure.

Environment-focused adaptation

The Esfahan Environment Organization is proposing to define a minimum flow requirement of $75 \times 10^6\,\mathrm{m^3/year}$ to the Gaw Khuny swamp to preserve the river and the swamp ecosystems in dry years. The impacts of these measures have shown that the pollution rates decreased in 2000, whereas in the same year the basin experienced a severe drought. The recent regulations have committed the industrial sector to increase their wastewater treatment efficiency, seen the installation of new wastewater treatment facilities and a number of factories have been relocated to other places. The painting units of the textile factories committed to shift to locations that are far from the river.

In spite of these environmentally friendly measures, results of the modelling framework show that for all of the future periods, BOD beyond the return flow of Esfahan water treatment will deteriorate. Even for May when discharges are highest, BOD is still higher than 10 mg/l. Another point that came out of the results of ZWAM is that even if no fixed amount of water was allocated to the swamp, at least 75×10^6 m³/year reached the swamp. This amount is obtained from return flows of the upstream demand sites. This was checked with historic records of inflows to the Gaw Khuny swamp, showing that inflow to the swamp ranged from 30 to 639×10^6 m³ over the last years. Yet in recent years (1997–2000) annual inflows have usually been down to 30×10^6 m³ as a result of the severe drought. The difference between this 30×10^6 m³ and the minimum flow requirement of 75×10^6 m³ should come from future additional inflows of Lenjan Tunnel and Tunnel No. 3 to the dam that will be exploited after 2010. So, even if no water is allocated for the swamp, it will still get a volume close to 75×10^6 m³.

Domestic- and industry-focused adaptation

The total amount of water presently allocated to the industrial sector is relatively small at about 100×10^6 m³, since there is not much industrial development in the basin that relies on water.

The only hydropower unit in the basin is the Chadegan Dam. The total electricity consumption in the basin has been estimated at about 3000 GWh. The Chadegan Dam produces only 5% of this amount. The main source of power in the basin is natural gas. Climate change can be considered as irrelevant for total power production.

The major measure that can be taken to reduce domestic demand is modification of the drinking water networks to reduce the present losses (25%, i.e. a reduction of 80 m³/year per capita to 60 m³/year per capita).

Combined adaptations

Period 2010–2039

If we consider the total demands for the period 2010–2039, we can see a rise from 2376 $\times 10^6$ m³ in year 2010 to 2522×10^6 m³ in 2039. As was pointed out earlier, two tunnels (Tunnel No. 3 and Lenjan Tunnel) are presently under construction and will be operational before 2010. This water diversion has been added to the present amount of available water. It is assumed that 425×10^6 m³ is the maximum capacity of these tunnels. Results show that when comparing A2 and B2 scenarios for this period, the basin will face more severe and longer water deficits in the future under climate change.

The strategies have been compared with the present water distribution and cropping patterns (Table 6.5). Applying the environment-focused adaptation strategy to

Table 6.5. Effect of the different adaptation strategies on the indicators.

	A2				B2			
	BAU	Env	Agr	Dom	BAU	Env	Agr	Dom
Food								
Total produced calories (10^9 kcal/year)	2489	2384	2968	3123	2373	2268	2824	2979
Max. shortage ($\times 10^6$ m^3)	600	670	487	592	835	865	682	592
No. dry years	15	20	10	6	15	15	14	14
Max. continuous dry years	7	7	3	2	15	15	14	14
Environment								
Inflow $<75\times 10^6$ m^3/year	12	0	0	0	15	0	0	0
$75<$inflow$<140\times 10^6$ m^3/year	18	12	12	12	15	15	15	15
Inflow$>140\times 10^6$ m^3/year	0	18	18	18	0	15	15	15

BAU, business as usual; Env, environment-focused adaptation; Agr, food adaptation (rice replaced by wheat); Dom, urban and industrial adaptation.

save the swamp causes a near 5% reduction in food production and higher water shortages. Applying the food adaptation strategy eliminates the previous negative impacts, increases food production up to 20% and reduces water shortage. For the next step, applying the domestic/industry adaptation, agricultural production may increase up to 25% compared to BAU and lower water shortage. It is evident that reduction in agricultural demands has a significant impact in reducing the vulnerability of the basin to water deficits.

Period 2070–2099

For the period 2070–2099, competition between domestic and agricultural demands will be more serious. While the present domestic demands are about 10% of available water resources, it rises to 25% at the end of this century.

The present water resources cannot meet the basin's agricultural and industrial water demands. In addition, an increase in domestic demand can be expected due to population growth. For this period, the basin will face more water shortage and scarcity. Similar to the previous periods, adaptation strategies have been examined for this period (Table 6.6).

As already pointed out, the Behesht Abad trans-basin project is one of the projects that is proposed by the Esfahan Water Authority to transfer water from the neighbouring basin (Karoon Basin). This project requires huge investments and may have some negative impacts on hydropower infrastructures in the Karoon Basin. More investigation is needed to assess the positive and negative aspects of this tunnel. The transfer capacity of the project is between 700 and 1000×10^6 m^3, but for the present analysis 700×10^6 m^3 has been assumed. Including this volume of water in the total amount of available water does not only reveal water shortage, but also generates new capacity to improve agricultural lands and produce more food. Such an improvement is definitely required, when the basin should accommodate and feed 7–9 million

Table 6.6. Similar to Table 6.5, but for the period 2070–2099. Tun is constructing the new Behesht Abad tunnel, which will transfer 700×10^6 m from the neighbouring Karoon Basin.

	A2				B2			
	BAU	Env	Agr	Tun	BAU	Env	Agr	Tun
Food								
Total produced calories (10^9 kcal/year)	2245	2140	2665	4557	2272	2167	2698	4590
Max. shortage ($\times 10^6$ m³)	778	848	665	0	1201	1271	1088	0
No. dry years	19	19	18	0	17	20	16	0
Max. continuous dry years	11	11	11	0	3	4	3	0
Environment								
Inflow $<75 \times 10^6$ m³/year	14	0	0	0	11	0	0	0
$75 <$ inflow $< 140 \times 10^6$ m³/year	16	14	14	14	18	11	11	11
Inflow $> 140 \times 10^6$ m³/year	0	16	16	16	1	18	18	18

people. While present resources of the basin are enough to produce 3460 kcal/day per capita, it will reduce to almost 810 kcal/day per capita in year 2099. Including the new tunnel, this will increase to 1400 kcal/day per capita. Thus, even after constructing the tunnel, the basin will still need to import a substantial amount of food.

Conclusions

In this study, climate change impacts and adaptation strategies relating to the water resources, food production and environmental preservation of the Zayandeh Rud Basin were assessed for two time periods, 2010–2039 and 2070–2099 by implementing GCM projections in a modelling framework. The results show a negative impact on the available water resources and a possible decline in water quality.

Rice, potatoes, wheat and barley were selected as the basin staple food to evaluate responses to future climate change. In general, crop production will increase because of a positive impact of enhanced CO_2. However, this increase is not enough to compensate for the negative impacts of a decline in water quantity and quality. Among the evaluated crops, rice shows the most negative responses in terms of average yield as well as yield variation. In the case of potatoes, the response is positive, although the yearly variation is estimated to be higher. Presently, domestic and industrial demands get 20% of the total available water, but at the end of this century this will reach about 35%. This increase is mainly a result of population growth. So, the proportion of agricultural water extractions is expected to go down.

To eliminate the aforementioned negative impacts, a number of adaptation strategies have been assessed and evaluated using a generic approach that can be used by policy makers and water resources managers. The following main conclusions and recommendations emerge from this study.

- Climate change will confront the basin with more severe water scarcity and salinity problems that makes proper water management at both basin and field level more crucial.
- The results show that there is a need for changes in cropping patterns. Specifically, rice will not be a recommendable crop.
- Pricing policy for crops can make growing crops with higher caloric values more appealing.
- Competition for domestic and agricultural water requirements is going to be more serious in the future. So population control is going to be an essential policy for the basin.
- The ecosystem of the Zayandeh Rud River suffers from domestic and industrial return flows. The treatment efficiency in these sectors should improve.
- The present water resources of the basin will not be sufficient for the future. Transfer of water from the neighbouring basins to the Zayandeh Rud Basin is an essential adaptation measure. The impact of such transfers on the source basins requires careful assessment.

References

Aerts, J.C.J.H., Lasage, R. and Droogers, P. (2003) *A Framework for Evaluating Adaptation Strategies.* IVM research report series, R03-08. Vrije Universiteit Amsterdam, The Netherlands.

Anonymous (1993) *Comprehensive Studies for Agriculture Development of the Zayandeh Rud and Ardestan Basins.* Yekom Consulting Engineers, Conclusions, Vol. 29. Ministry of Agriculture, Iran.

ASCE (2000) Artificial neural networks in hydrology, II: hydrology application. *Journal of Hydrological Engineering* 5, 124–137.

Coulibaly, P., Anctil, F. and Bobee, B. (2000) Daily reservoir inflow forecasting using artificial neural networks with stopped training approach. *Journal of Hydrology* 230, 244–257.

Dawson, C. and Wilby, R. (1998) An artificial neural network approach to rainfall-runoff modeling. *Hydrologic Science Journal* 43, 47–66.

IPCC (2000) Climate change 2001: impacts, adaptation and vulnerability. Available at: http://www.ipcc.ch/pub/tar/wg2/index.htm

Koochaki, A. and Mahallati, M.N. (2001) A simulation study for growth, phenology and yield of wheat cultivars under the doubled CO_2 concentration in Mashhad conditions. *Journal of the Research Center for Deserts and Arid Zones* 6, 34–46.

Moeinian, M.T. (2000) The impact of drought on ecological factors of the Gaw Khuni swamp. Esfahan Environment Authority.

Morrison, J., Quick, M.C. and Foreman, M.G.G. (2000) Climate change in the Forest River watershed: flow and temperature projections. *Journal of Hydrology* 263, 230–244.

Parry, M.L., Rosenzweig, C., Iglesias, A., Fischer, G. and Livermore, M. (1999) Climate change and world food security: a new assessment. *Global Environment Change* 9, 51–67.

Ramsar (1975) The Ramsar convention on wetlands. Available at: http://www.ramsar.org

Sanjikumar, N. and Thandaveswara, B.S. (1999) A non-linear rainfall-runoff model using artificial neural networks. *Journal of Hydrology* 216, 32–55.

van Dam, J.C., Huygen, J., Wesseling, J.G., Feddes, R.A., Kabat, P., van Walsum, P.E.V., Greoenendijk, P. and van Diepen, C.A. (1997) Theory of SWAP Version 2.0. Technical Document 45, Wageningen Agriculture University, Wageningen, The Netherlands.

WEAP (2002) Water Evaluation and Planning System. Available at: http://www.seib.org/weap

7 Increasing Climate Variability in the Rhine Basin: Business as Usual?

HAN KLEIN,[1] KLAAS JAN DOUBEN,[1] WILLEM VAN DEURSEN[2]
AND ERIK DE RUYTER VAN STEVENINCK[1]

[1]UNESCO-IHE Institute for Water Education, Delft, The Netherlands;
[2]Carthago Consultancy, Rotterdam, The Netherlands

Introduction

The River Rhine flows from the Swiss Alps through Germany, France and The Netherlands into the North Sea (Fig. 7.1). With a length of 1320 km, the River Rhine is the longest river in Western Europe with a total catchment area of about 185,000 km². The catchment has a population of approximately 50 million.

Fig. 7.1. The Rhine Basin and its countries.

As a result of industrial development and population growth, water demands increased for a large range of functions such as domestic water supply, industry, agriculture, hydropower generation, recreation, navigation and cooling purposes. In view of the functions and uses, the Rhine has undergone many modifications and has turned into a heavily engineered river. In the past the river was considered wild and it took more than 100 years to turn it into a navigable river, to canalize it, to equip it with power stations in the Alsace Plain and to make it the main water supply in the lower reaches. Worldwide, the Rhine is among the inland waterways with the highest traffic density, with the port of Rotterdam being the largest seaport and Duisburg one of the largest inland ports in the world (CHR, 1993).

Water use and agriculture in the Rhine Basin

Under present-day conditions, about half of the total area of the Rhine Basin is used for agriculture, approximately one-third is covered with forest, 11% is built-on area and the remainder consists of surface water. Agricultural land use is an important element in the rainfall–runoff relations in the Rhine Basin. Investments in land drainage, external water supply and large-scale use of biocides and fertilizers have increased agricultural production. Yet, at only 10% of total extraction, agricultural water use is relatively low compared to that in arid climates. In this context, Dutch water management practices use a large amount of Rhine water to maintain a constant (ground) water level in the polders and low-lying areas of the delta in the summer.

Another important water use is navigation, which requires minimum water depths and is limited by maximum flow velocities. Hence, the Rhine, its tributaries and branches have been partially canalized and divided into sectors, separated by locks. Use of Rhine water for human consumption varies between 5% and 15% of the mean annual discharge. This may not seem much, but consumption is not spread evenly over the year: the demand is greatest in summer, when the mean discharge is lowest. This means that the available water must be carefully apportioned during dry summers (Speafico and Kienholz, 1996; RIZA, 2000).

About 20 million people depend on Rhine water as a source of drinking water (Dieperink, 1997). Most of them live in Germany and in The Netherlands. In the western part of The Netherlands, groundwater is often too brackish to be used as drinking water.

Threats to the water resources system

In recent years (1988, 1993, 1995 and 1998) peak flows and even small flood events caused substantial socio-economic problems, especially in the lower parts of Germany and The Netherlands. The encroachment of urban and industrial expansions along the river, together with agricultural developments, makes it impossible for the river to inundate the original valley. Approximately 85% of the floodplains have been withdrawn from the natural river. Hence the risk of flooding has increased, especially in downstream areas.

The Rhine has been subjected to enormous amounts of pollutants over a long period of time. Pollution reached its maximum between 1960 and 1975, when the Rhine was described as an open sewer. In particular, high loads of organic waste (sewage) resulted in oxygen levels too low for many fish species. The Rhine became a dead river, losing its function as provider of drinking water and depositing large amounts of polluted sediments in the river's tidal areas and on its floodplains. The situation improved significantly due to the installation of water treatment works by industries and cities during the 1970s and 1980s. However, accidents can still take their toll. A fire at Sandoz Chemical Industry in Basel, Switzerland, in 1986 led to severe pollution of the Rhine, resulting in massive death of macrofauna and fish.

The different water demands and uses often compete and impose stress on the water resource system in terms of human security and nature values. Increasing demands have led to a range of measures to control water resources. Specific measures concern drinking water supply and water pollution control, flood protection, energy production, navigation and more recently the ecological rehabilitation of the river. The relatively large number of countries, however, makes it difficulty to develop a common vision to cope with developments that may impact the trans-boundary water resources system of the basin. The pollution events of the 1970s and 1980s gave an impulse to the already existing International Commission for the Protection of the Rhine (IPCR), which has the commitment of all the important tributary countries. This Commission prepares plans for further improvement of the water and ecological quality of the river. It developed the Rhine Action Programme, which has a reduction of (sediment) pollution as one of its key objectives, and 'Rhine 2020', which focuses on flood protection.

Climate Change Projections

The Rhine originates in Switzerland as a mountain river, fed by glacier water, snowmelt and rainfall. As a result, the hydrograph upstream of Rheinfelden (the Swiss part of the Rhine catchment) is largely determined by high discharges in spring due to snowmelt. Between Rheinfelden and Lobith (where the Rhine enters The Netherlands), the Rhine changes character from a mountain river, though various stages typical for medium high mountains, to a lowland river. Elevation in this region is between 100 and 1500 m, and the current climate ranges from alpine towards the moderate humid climate of the North Sea region. Inflow into the river in this stretch is determined by a large precipitation surplus in winter and a precipitation deficit during summer. On crossing the Dutch–German border at Lobith, the river has become a typical lowland river. Here its hydrograph is determined equally by the snowfall and snowmelt regime of the Alps and by the seasonal effects of precipitation surplus.

The SRES HA-2 scenario (see Chapter 2) simulates an increase in average temperature for the entire catchment upstream from Lobith (for the location see Fig. 7.2) from 8°C in the current situation to 9°C for the period 2010–2039 and 12°C for the period 2070–2099. For the catchment upstream from Rheinfelden (Fig. 7.2), the yearly temperature increase ranges from 6°C (current situation) to 7°C (2010–2039) to 10°C (2070–2099) (Fig. 7.3).

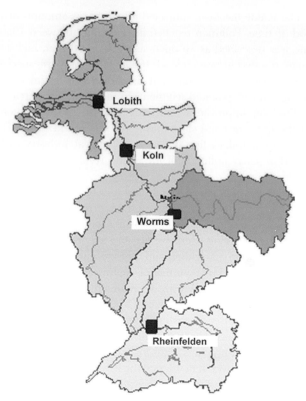

Fig. 7.2. Locations of Lobith and Rheinfelden stations.

The calculated precipitation changes are given in Fig. 7.4. For the entire basin above Lobith, average yearly precipitation will change from 2.6 mm/day (current situation) to 2.7 mm/day (2010–2039) and 2.6 mm/day (2070–2099). For the area above Rheinfelden, the projections are 4.1 mm/day (current situation), 4.2 mm/day (2010–2039) and 4.1 mm/day (2070–2099). Hence, the changes in *annual* water availability in the Rhine catchment will be relatively small. The *temporal* distribution through the year, however, is expected to change significantly. It is projected that the Rhine will change from a combined snowmelt–rain-fed river to an almost entirely rain-fed river. Precipitation projections show a decrease in runoff in summer and an increase in winter and spring. Although the actual precipitation in summer is expected to increase, the accompanying increase in evapotranspiration will result in a net decrease in effective precipitation in summer and therefore in a reduction in river discharges. Such a change will mainly affect areas that already are sensitive to drought. Additional consumption of water for irrigation may create additional depletion of water reservoirs and groundwater. The frequency of peak-flow events may increase the number of flood events in the downstream part of the river (Kwadijk, 1993; Grabs, 1997; Middelkoop *et al.*, 2000, 2001).

Figure 7.5 (graph for Lobith) shows the variability in precipitation for the entire

Lobith

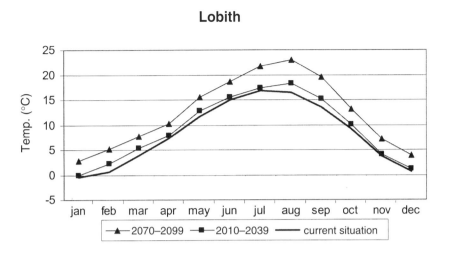

Rheinfelden

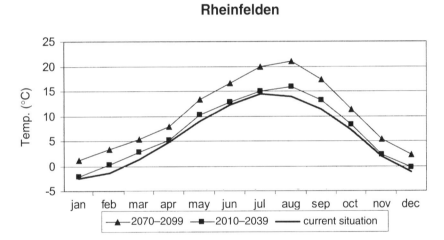

Fig. 7.3. Temperature in the SRES HA-2 scenario for a downstream and upstream station (Lobith and Rheinfelden).

catchment and the catchment upstream from Rheinfelden. Variability in precipitation for each month is here defined as the 95th percentile of the monthly precipitation for the analysed period minus the 5th percentile of the monthly precipitation. The figure shows that the variability in precipitation increases largely under the climate change projections.

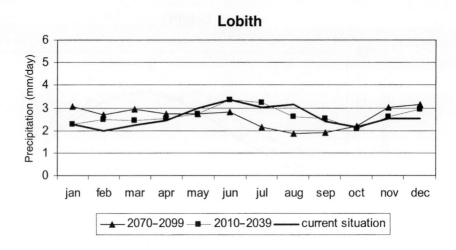

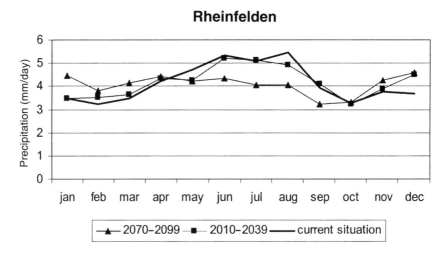

Fig. 7.4. Precipitation for the SRES HA-2 scenario (Lobith and Rheinfelden).

Impacts from Climate Change

Hydrology

The SRES scenarios used in this study result in a decrease of summer discharge and an increase in winter discharge. The results of the simulations of the SRES HA-2 scenarios with the Rhineflow model are given in Fig. 7.6 with the 10th and 90th percentile runoff intervals. A clear change in mean runoff is visible. The increased runoff in winter, when soils are saturated, is mainly a result of changing temperatures and increased precipitation, which greatly affects the mechanism of snowfall–snowmelt. The decreased summer runoff is caused by increased temperature and related

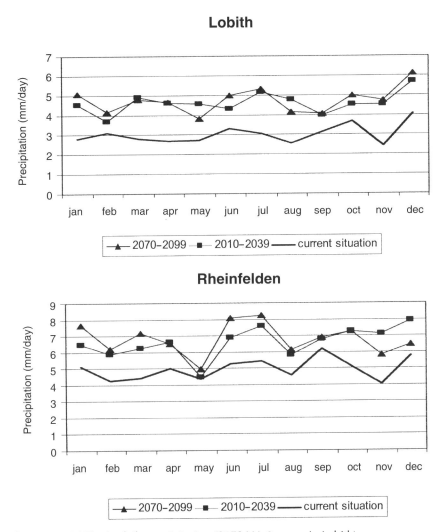

Fig. 7.5. Variability in daily precipitation (SRES HA-2 scenario Lobith).

increased evapotranspiration and by reduced contributions from snowmelt. In addition, Fig. 7.6 shows a larger variability in runoff, illustrated by the broadening of the variability bands around the mean runoff in the projection periods.

An increase in precipitation may lead to increased peak flows and hence increased flood risks in winter. The increased temperatures in summer could lead to higher local precipitation extremes and associated flood risks in small catchment areas (Kwadijk, 1993).

The projected changes in climate and hydrology will affect various user functions of the River Rhine. Higher average and extreme temperatures will enhance the demand for freshwater, in particular for agriculture and direct human consumption. Changes in precipitation patterns, particularly over regions that already are sensitive,

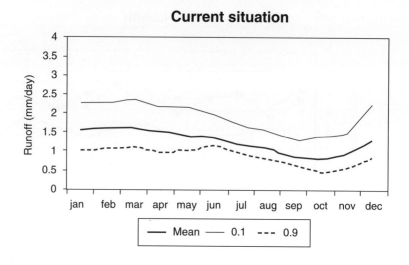

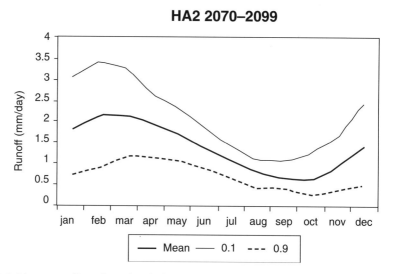

Fig. 7.6. Mean runoff combined with the 10th and 90th percentile of runoff distribution at Lobith for the current situation and projected according to the SRES HA-2 scenario, 2070–2099.

may lead to increased demand for water for irrigation purposes, especially for soils with low water-storage capacities. The Netherlands could face the desiccation of most of its wetland areas or be forced increasingly to rely on the Rhine to maintain present water levels. Especially during periods of low river discharge, groundwater aquifers could be affected by increased saltwater intrusion as sea level rises. Given a possible 4°C increase in temperature, and a rise in the alpine snow line of 500–700 m in the summer, the flow of the Rhine could decline by 10% during this season. Below are the major expected impacts of climate change on the human environment, ecosystems, food security and industry (see also NWP, 2003).

The environment

The impacts and effects of extreme peak flows can be enormous. Flooding and subsequent damage and losses will occur along river stretches where design criteria for flood protection are exceeded. Historical flood records in Germany and The Netherlands illustrate that natural disasters can devastate prosperous (local) societies and economies for a long period of time.

Both high and low discharges can result in damage to habitats. Peak flows can result in disturbance and flushing of macrofauna and pollution of groundwater. Low discharges will have consequences for the availability of water for drinking, irrigation and nature, for example by lowering groundwater tables. On the other hand, a more dynamic river flow could also result in a more diverse environment, stimulating biodiversity.

Assuming that the total loads of pollutants via point and diffuse sources will remain the same, concentrations of dissolved substances will increase during low water flows. Although some reduction of the total pollutant load may be realized, concentrations of pollutants may reach levels that are toxic for river organisms and will cause problems for the supply of drinking and irrigation water. Increased levels of organic compounds, including algal biomass, might cause low oxygen levels and kill river fauna. An increase in temperature can amplify this effect. An increase of water temperature might also affect the abundance and diversity of riverine species (disappearance of sensitive species, colonization by exotic species). Finally, increased intrusion of salt water during low discharges can impair water quality in downstream stretches of the river.

Food security

The relationship between the changed discharge regime of the River Rhine and food production in the countries bordering the Rhine is complex. In the Rhine Basin in general, the importance of river water for irrigation is much less than in the arid regions of the world. However, an increased risk of salinization in the downstream part of the Rhine Basin will be of major concern in periods of reduced flow, as this will limit the use of Rhine water for irrigation.

In the EU agriculture accounts for only a small part of the GDP. Moreover, the Rhine catchment already suffers from huge overcapacity of agriculture, and subsidies are devoted to controlling overproduction. Therefore the vulnerability of the overall economy to changes that affect agriculture is low. The aim of the present EU agriculture policy is to focus more on limitation than on intensification of agriculture, with a shift towards sustainable production. This will be accomplished by taking marginal lands out of production and the stimulation of organic farming (European Commission, 2003). In general, the effects of climate change (increase in CO_2, rise in average temperature and a net increase in precipitation) in principle are expected to have a positive effect on crop production. An increased risk and probability of water stress in late summer and early autumn will be compensated largely by an earlier start of the growing season.

Industry

Water is an important agent for cooling processes in electricity production (both nuclear and conventional production of electricity) and in other heavy industries. The major industrial regions in Germany in particular depend very much on the hydrological regime of the River Rhine. Increased drought during the summer seasons, as well as increased water temperatures, might seriously hamper these industrial processes.

Inland navigation plays a major role in bulk transport and is becoming increasingly important for container transport. Low water levels result in unreliable transport means, because of the link with the allowable draught of the vessels. Peak flows will hamper navigation because high water levels can also result in navigation restrictions.

Adaptation Strategies

Much work on adaptation for the Rhine Basin has been conducted, as outlined in the Non-structural Flood Plain Management study of the ICPR (2002). The main adaptation types that address the impacts described in the previous paragraphs can be arranged in three categories: (i) land use modification; (ii) structural measures; and (iii) non-structural measures.

Adaptation categories in the Rhine Basin

Land use modification

Land use modification measures first of all address increased flood and drought risks. By changing land use practices, people and society can adjust to the total amount, as well as the variability in the amount, of water available. Reduction of the vulnerability of natural water systems to extreme conditions is a guiding principle in this approach. For example, in the case of droughts, catchments with significant artificial storage in reservoirs, groundwater recharge or regulated flows can withstand drought for a longer period. Or, in the case of increased floods, planned urban developments in floodplains should be reconsidered and at least be designed to limit potential flood damage to an acceptable minimum.

With a (potential) increase in flood frequency and flood levels, the duration of floods and the extent of flooding will increase. As a result, the area of flood-prone land will expand and the economic value of already flood-prone agricultural areas will be reduced. It is expected that flood-prone agricultural areas will increasingly be used for nature rehabilitation or reforestation projects. This shift in land use can be combined with recreational use, although the main role often is a buffering function for excess runoff. Limiting agricultural activities in floodplains as an adaptation to increasing discharges is expected to have a positive impact on the environment. Autonomous development of nature in these floodplains is expected to contribute to a higher diversity of habitats, which in turn will create conditions for (re)colonization by species that have disappeared or are (locally) threatened. An additional positive impact on nature will be the reduced application of fertilizers and pesticides when agricultural activities in floodplains are abandoned. However, pollutants from upstream sources and resuspended polluted sediments might settle in the new nature areas and interfere with the full development of these areas.

Afforestation strategies aiming at (amongst others) a higher retention of the catchment will have a (potentially large) effect on groundwater storage in the catchment. This will allow urban water industries to rely more on groundwater as their main source of water. However, this trend is completely neutralized by the current policy to shift the source of the urban water production from groundwater to surface water.

With regard to energy use and energy production, an adaptation strategy that is very much linked to industries is the production and use of energy crops. In Western European society, where overproduction of food crops is becoming a major economic and environmental problem, the production of energy crops might prove a contribution to the solution. Whether this makes the energy industries less dependent on water for energy crop growth and cooling water demands remains to be seen (Stephens *et al.*, 2001).

Structural adaptation

Structural adaptation includes raising dykes or developing flood emergency storage areas. Strengthening and raising dykes to cope with floods will affect the river landscape in a negative way. No positive impact is foreseen for nature. When dykes are relocated to increase the width of the floodplain, this will benefit the environment, provided the floodplains are left for nature development. During the construction phases measures have to be taken to limit environmental impacts. Dredging to deepen the riverbed is another adaptation option, but it is expected to have negative impacts on the environment as it may destroy valuable habitats for riverine species and lower groundwater tables locally. Moreover, deposition of (polluted) sediments may lead to pollution. Increasing storage capacity by the creation of retention basins offers the possibility to actively contribute to the construction of areas with the potential for realizing high environmental nature values.

A structural measure in agriculture is to introduce new crops. In view of the constraints and risks possibly resulting from climate change, a move towards the development and introduction of more resilient crops and crop varieties is to be expected. At the same time the use of perennial crops (e.g. fruit and grapes) can provide a buffer to reduce the impact of short-term variability, although in general an increase in yield variability would be expected. Energy crops such as willow that are only harvested every few years would provide the best insurance against impacts of short-term climatic fluctuations. Use of such crops on the other hand has implications for the hydrology of an area. The soil moisture is utilized to a higher degree by deeper roots, resulting in a higher degree of depletion in the unsaturated zone during the growing season, and a higher storage capacity at the beginning of the winter season.

For navigation the major measures are dredging and structural measures to maintain the water depths in the navigation channels. A future with climate change will increase the number of sluices and weirs to maintain water levels and will increase the attention towards the innovation of technical water management solutions.

To secure water availability for industrial and cooling purposes there is already a trend towards the construction of larger reservoirs. These reservoirs can be actual structures, such as the three reservoirs of the Waterwinningsbedrijf Brabantse Biesbosch, serving 5 million people in the south-western part of The Netherlands, or can be provisions to increase groundwater storage.

Non-structural adaptation

As mentioned before, the EU agricultural policy is the dominating factor determining the present situation and future developments of agriculture in the Rhine Basin. This includes the present system of subsidies. On top of that, national legislation and regulating measures have a significant role.

Crop selection and risk insurance (apart from excluding climate variability through greenhouses, stables, etc.) are important adaptation measures. Research on new varieties and practices by government and private research establishments and dissemination of results through extension services and professional associations is well established and can easily incorporate climate change issues. The driving force will be adaptation to economic developments rather than climate change, although the risk assessment aspect ('commercial risk') will incorporate the aspect of increase in climate variability.

Insurance mainly focuses on the impacts of extreme events (floods due to unusual rainfall, hailstorms, wind damage, etc.). An increase in frequency and extent of these events can make insurance cover unavailable or unaffordable. Relief measures take the place of insurance cover in case of disasters. (Partial) compensation of damage is a common response to floods, hurricanes and other natural extreme events. The reliance on relief measures and insurance cover is to a certain extent counteracting the implementation of adaptation measures, since it reduces the perceived (financial) risk.

Non-structural measures to deal with climatic variability and extremes can be relatively simple and cheap. An example is the use of shelter-belts to protect orchards against strong wind. This practice could be expanded to additional areas and crops to deal with increased risk of storms through amplified climate variability. A spin-off could be amelioration or improvement of local climate in semi-urban areas.

Non-structural strategies for navigation include all measures that are related to the composition of the fleet and the design of individual ships. Innovative ship design (such as the so-called river snake design) can decrease the draught of the ships, while maintaining the transport load. New materials can result in lighter ships, decreasing draught per transported unit of load. Other strategies might include the use of multi-modal transport means (using combined ship–truck–train transport chains to become less dependent on the available water depth) or strategic reservoirs at the destination location.

Non-structural strategies for industries include relocation of the production processes to become less vulnerable to floods and low water-level situations, and protection of water sources such as groundwater. This also includes the storage of raw materials and of waste to avoid water pollution during floods and to ensure continued production during times of interrupted supply. Increased water use efficiency (e.g. reuse of process water and cooling water) will contribute to reduce the dependency on external water supplies. The latter is particularly relevant during periods of limited water quantities (e.g. prolonged dry periods) but would also play a role when water quality becomes the limiting factor. This can occur both during very low discharges in the rivers or during very high flows.

Institutional arrangements

The International Commission for the Protection of the Rhine (IPCR) was initiated in the 1950s following concerns about pollution of the river and the implications for drinking water supply in the downstream regions. The IPCR has representatives from Switzerland, France, Luxembourg, Germany and The Netherlands. The ministers of the Rhine states met for the first time in 1972. They decided to draft international conventions and programmes to combat pollution. In 1976 the ministers adopted two conventions and the long-term working programme to reduce pollution. The Sandoz chemical spill after a fire in the factory and the subsequent large-scale fish kills in the Rhine downstream of Basel triggered a transition from the formulation of pollution control agreements to the implementation of the Rhine Action Programme for Ecological Rehabilitation (RAP) in 1987. At that time flood protection and flow management did not yet play a major role. After the flood events of the 1990s, the Conference of Rhine Ministers in 2001 adopted the Rhine 2020 programme on sustainable development of the Rhine. This succeeds the RAP, but also includes flood prevention and protection. The EU directive 2000/60, establishing a framework for EU action in the field of water policy (WFD), will contribute to the implementation of the 'Rhine 2020' programme. Important targets of the Rhine 2020 programme include ecosystem improvement, flood prevention and protection, water quality improvement and groundwater protection.

Adaptation strategies

On the basis of the discussion above, a number of key adaptation measures have been identified and combined into five strategies. Three adaptation strategies are dedicated to cope with impacts on the environment, food security and industrial capacity, respectively (Table 7.1). Furthermore, an integrated strategy combines the most important element of these three strategies. Finally, a BAU (Business As Usual) strategy is used for comparing the performance of the adaptation strategies. The principles of the policy plan 'Water Management for the 21st Century' (WB21, 2000) have been used to develop the adaptation strategies. For general background information see also the report of the 'Dutch Dialogue on Climate and Water' (NWP, 2003).

The adaptation strategies are defined as follows.

1. 'Environmental strategy': This strategy consists of the modification of land use practices; floodplain developments resulting in nature development and rehabilitation; implementation of upstream retention areas with nature development (reservoirs/wetlands) and afforestation plans; floodplain rejuvenation; the reduction of discharges and loads of domestic and industrial wastes; the reduction of sewer overflows and cooling water use.

2. 'Safety strategy': This strategy consists of the strengthening and heightening of dykes and levees; the development of emergency storage areas along downstream river stretches; the improvement of water retention in upstream parts of the catchment; enhancement of rapid discharge of floodwaters in downstream river stretches; raising awareness and preparedness by improved flood forecasting and information systems; inclusion of risk assessment in spatial planning; the provision of insurance systems and emergency plans.

3. 'Industrial strategy': This strategy includes the deepening of navigation channels; the implementation of additional weirs and sluices; lengthening of groynes;

Table 7.1. Adaptation measures and strategies.

Measures	Strategy		
	Environment	Safety	Industry
Land use			
Floodplain storage and retention	■	■	
Afforestation of catchment	■	□	
Growing energy crops	□	□	■
Structural			
Dredging		■	■
Sluices/weirs/groynes			■
Reservoirs/bypasses	■	■	■
Heightening/enforcing embankments		■	□
Non-structural			
Forecasting		■	□
Improving preparedness		■	□
Adapt ships and transportation strategies			■
Improve water use efficiency, reduce discharge of pollutants	■		■
Incorporate risks in planning	□	■	□

■, Core measure of the strategy. □, Measure probably to be implemented but not essential for the strategy.

implementation of cooling water reservoirs (upstream or local) and towers; the implementation of alternative energy sources (wind, solar, etc.).

4. 'Integrated strategy': Implementation of retaining, storage and discharge measures and increasing awareness, preparedness and forecasting facilities. In the framework of this report this strategy is assumed to include the strengthening and heightening of dykes and levees in the lowland region of the River Rhine.

5. 'Business as usual/No adaptation': Based on autonomous developments. The crop yields are expected to increase, but total production will decrease by reduction of the area used for agriculture (by abandoning marginal lands vulnerable to the effects of climate variability and especially by the EU policy measures). Due to this, some nature development in the floodplain will take place. Otherwise the effects of the strategy will be neutral or negative. The latter especially concerns major economic and safety aspects and this strategy is therefore not considered a realistic option.

Evaluation

The qualitative evaluation of the adaptation strategies is based on expert judgement using a number of indicators, which have been selected to represent various impacts that can be expected. A brief description of these indicators and some considerations concerning key adaptation aspects is given below. Table 7.2 illustrates an assessment matrix of the five adaptation strategies.

People at risk

Flood protection measures aim to reduce the number of people at risk. The most effective way is to reduce peak flows by upstream storage and efficient discharge. Traditional measures of dyke strengthening and heightening will on one hand reduce the frequency of flooding but at the same time potentially increase the magnitude of a flooding event (depth and extent). The flood protection measures in the 'Safety' strategy will reduce risk in the short term (2039) but possibly increase risk in the long term (2099), especially since the frequency and magnitude of extreme events will increase during that period.

Economic losses

Economic losses are defined as direct and indirect damage to property and income. Without protective measures (flood proofing, etc.) even relatively small events (in terms of flood depth, extent and duration) can result in considerable direct and indirect damage. Typical examples are the pollution by fuel oil from flooded oil tanks for household heating situated in the basement of buildings and damage to electrical and other equipment in the basement or ground floor of private, commercial or public buildings (e.g. as occurred in Dresden in 2002; www.dw-world.de). Possible interruptions of normal trade through blockage or disruption of communication and transport will result in considerable economic losses as well. The latter can be the result of floods and extreme weather events, but also of very high or very low flows in the river that interfere with shipping.

Fish diversity

The main impact on fish diversity will be the loss of habitat due to further canalization of the river in the 'Industrial' strategy. Introduced species may profit from new conditions directly through a more favourable environment (e.g. temperature) or through a reduced resilience of native species. In general the risk of introduction of new species will be a determining factor, although some 'pockets' of special conditions will allow the continued survival and potential spreading of deliberately or accidentally introduced exotic species (e.g. local population of tropical fish *Poecilia reticulata* in cooling water of power plants in Ijmuiden, The Netherlands).

Upstream forest

One of the measures to influence peak flow levels is to increase the upstream retention of rainwater. In the Rhine Basin this is of special concern in periods when the evapotranspiration of agricultural crops is low.

Table 7.2. Expected effect of the five strategy options in the period up to 2039 and the period up to 2099.

	Strategy									
	No Adaptation		Environment		Industry		Safety		Integrated	
Indicators	2039	2099	2039	2099	2039	2099	2039	2099	2039	2099
Human										
People at risk (no.)	−	− −	+ +	+	0	0	+	+(−)	+	+ +
Economic losses	−	− −	+	+	+	+ +	+	0/−	+ +	+ +
Ecosystem related										
Fish diversity	0	0	+	+	−	−	0	0	+	+
Upstream forest (ha)	0	0	+	+ +	0	0	0	0	+	+
Floodplain forest and wetlands (ha)	0/+	+	+	+ +	0	0	0	0	+	+
BOD	0	0	+/−	+/−	0	0	0	0	0/+	0/+
Fertilizer	0	0	0	+	0	0	0	0	0	0/+
Lateral and longitudinal freedom (%)	0	0/+	+	+	− −	− −	0	0	+	+
Food production										
Total production	0	0	−	−	0	0	0	0	−	−
Average yield	+	+	+ +	+ +	+	+	+	+	+	+
Variation in yield	−	−	+/−	+/−	−	−	−	−	0/−	0/−
Food security										
Variation in farm income	−	−	0/−	0/−	−	−	−	−	−/0	−/0
Water productivity	0	0	−	−	0	0	0	0	0/−	0/−
Industry										
Weirs (no.)	0	0	0	0	+	+	0	0	0	0
kW produced (hydropower and cooling water)	−	− −	0/−	0	+	+ +	0	0	0	0
Un-navigable days (no.)	−	− −	0	0/+	+	+	0	0	0	0/+

+, positive effect of the strategy on this indicator (quantitative effect can be an increase or a decrease!); −, negative effect of the strategy on the indicator; 0, no effect of the strategy on this indicator.
+/−: ..
++: ...

More extensive land use will help to improve infiltration, while (evergreen) trees will maintain evapotranspiration and also will intercept a considerable amount of water. An increase in upland forest is therefore a management measure as well as an element of the Environmental strategy.

Floodplain forest and wetland area

An increase in wetland and floodplain areas is considered as an environmental benefit, since it could attract more species and enrich existing ecosystems. Increasing vegetation densities in the floodplains has an adverse impact on flood risks by increasing roughness and hydraulic resistance, thereby reducing the conveyance capacity during peak flows.

BOD

A reduction in pollution loads will in principle result in an improved water quality. If flows are reduced at the same time (due to climate change effects and/or by upstream retention/utilization), the actual water quality could still deteriorate by lack of dilution.

Fertilizer

The modifications in land use to increase water retention and groundwater infiltration in the upstream parts of the catchment will result in a reduction of the area used for agriculture and therefore in a reduction in the total amount of fertilizer used.

Lateral and longitudinal freedom

Land use changes due to floodplain development and retention areas will positively affect the degree of lateral freedom in the river system and could contribute to the (re)establishment of 'ecological connection zones'. Impacts on longitudinal freedom would only result from the additional weirs that are part of the 'Industry' strategy. These impacts could partially be mitigated by the inclusion of fish ladders in the design of the weirs.

Total production

As a result of the increased CO_2 levels, combined with an increase in temperature and net increased precipitation, the average yield of crops is expected to increase considerably. As such this is independent of any adaptation intervention. The overall result of the developments in agriculture is that areas with a reduced suitability (regarding soils, slope, exposure, etc.) are taken out of production.

Average yield

As mentioned above, the average yield of crops is likely to increase as a result of potentially more favourable climate conditions in combination with improved agricultural practices. In the 'Environment' scenario, it is assumed that the more marginal agricultural areas will be used for nature development. Concentration of agriculture in the most suitable locations will have an additional positive effect on average yield (per unit area).

Variation in yield

Although the increased climate variability is a factor that will add to the yield variation, the concentration of agricultural production in the most suitable areas will help to mitigate yield variation.

Variation in farm income

This aspect is directly linked to total production and yield variation but would also depend on the (new) role of farmers in nature conservation, management and recreation. This will ameliorate the negative impacts of climate variability on farm income, at least for some of the farms in some parts of the catchment.

Water productivity

Climate change in the Rhine Basin is associated with an increase in rainfall, together with an increase in evapotranspiration as a result of the rise in average temperature. Although (biomass) productivity will increase, this will be associated with a more or less equivalent increase in evapotranspiration. Water productivity as kg/m^3 will therefore be more or less constant (assuming no or only a limited water stress during the growing season).

Weirs

If water depth becomes a limiting factor for navigation, the construction of additional weirs would serve to maintain the necessary depth in critical river stretches. It has to be noted that the reduced capacity of the fairway may be a more critical factor than the actual water depth. This concerns in particular the waiting times at shipping locks needed when weirs are constructed and reduced passage heights under bridges with changing peak flow regimes.

kW produced

For the production of electricity, the discharge of the river can be a limiting factor. The discharge determines the capacity of hydropower production. In the River Rhine the only major hydropower plants are located in Switzerland. The thermal power plants in the downstream areas of the Rhine depend on the river as a source of cooling water (and the transport of coal). Adaptations to times of low flow are the maintenance of the minimum base flow through upstream storage or the development of cooling water reservoirs. A reduction of the need for cooling water can be achieved by constructing cooling towers or by the use of alternative energy sources (solar, wind) without cooling requirements. Energy crops are not part of the solution of the problem, as they would be used as fuel in thermal power plants (requiring cooling water).

Un-navigable days

Both during periods of peak flows and periods of low flows, navigation on the Rhine and its tributaries will be impeded. This can be caused by physical limitations or regulatory constraints. An increase in extreme high or low flows will increase the frequency of reduced navigability. Most strategies aim at reducing peak flows by upstream storage, allowing release of water during periods of low flows. Navigation will benefit to a certain (limited) extent from this levelling of extremes, but only the specific measures in the Industry scenario will have a positive effect.

The measures in the Industry strategy prevent economic losses from reduced working days and reduced capacity. The projected economic losses would be higher due to a combination of increased climate variability and climate change trends. By implementing the specified measures, economic benefits increase after all. Increasing the heights of dykes and levees in the Safety strategy implies an increased protection factor, but the number of people potentially at risk and the magnitude of potential damage will increase as well. The use of emergency storage areas could result in considerable economic losses. This is more likely to occur in the later part of the century. In that case the economic benefits of protection would be partly eliminated. With an increasing risk factor, combined with a growing population and ongoing investments in flood-prone areas, the benefits of protective measures will increase, although the measures themselves may not be changing.

Due to a more sustainable approach, the Integrated strategy is expected to have a positive effect on the economy. Increasing base flows in summer will enhance dilution, although the pollution level (load) itself is not affected by this strategy. The increased forest area will reduce the amount of fertilizer used, but due to the soil reservoir this is only a long-term effect. Retention measures upstream, combined with appropriate operation schemes, will result in a higher base flow in summer. This will be especially important in the latter half of the century and the forestation may only be required after the 2039 period.

Finally, maintenance of flows for (hydro)power production could also benefit navigation.

Preferable strategy and key indicators

Obviously the number of people at risk and potential economic losses are two core indicators in the evaluation of the five strategies. Given the number of people at risk, it is perhaps surprising that the safety strategy has not been allocated the highest score. The main reason for this is that the amelioration of extreme flows in the Environment

Table 7.3. Quantitative comparison of scenarios against the five adaptation strategies.

Strategy	2039	2099
BAU	−4.5	−7.5
Environment	6	9
Industry	0	2
Safety	1	−1.5
Integrated	6	8

and Integrated strategies, combined with a limitation or even reduction of the population in flood-prone areas for ecosystem development, results in the highest score for these two strategies. Economic losses are reduced by the Industry strategy, but here also the Environment and the Integrated strategy score nearly as well, or even better.

The 'natural' condition of the floodplain is best described by the indicator 'lateral and longitudinal freedom'. Additional weirs and canalization in the Industry scenario obviously will have a clear negative effect, while the Safety scenario does not affect this indicator, as this scenario does not include significant interventions in the floodplain itself.

The indicators kW produced and number of un-navigable days are specific for the industrial interests. Only the specific measures in the Industry scenario have a positive effect. In the other cases the result is mostly neutral, although again the reduction of the duration and magnitude of the highest and lowest flows in the Environment and Integrated scenarios will have some influence.

As discussed earlier, the productivity of the agricultural sector is predominantly determined by political considerations at the EU level. In the evaluation of the climate adaptation scenarios the food production and food security indicators therefore do not play an important role.

By allocating values to the indicators scores in Table 7.3 a quantitative comparison can be made, although it has to be stressed that in this case it has been assumed that all indicators have an equal weight. In practice the selection of a strategy would depend to a large extent on the weighing of the different impacts.

Table 7.3 shows that the BAU strategy has very undesirable results, which will increase after 2039. The opposite applies for the Environment and the Integrated strategy, whereas the Industry and Safety strategy have a low score, with a negative score for the Safety strategy in the long term. The reason for this low score is partly due to their limited scope with hardly any positive spin-off in the other sectors.

The difference between the Environment and Integrated strategy is mainly due to differences in the agricultural sector, while the economic losses are lower in the Integrated strategy. Taking into account that socio-economic considerations will be of major importance, an integrated strategy is the most likely choice.

Cost and benefits

Apart from estimates of maximum potential damage, it is difficult to compare the financial benefits of the different strategies. The (Dutch) Emergency Flooding Areas

Commission (Commissie Noodoverloopgebieden, 2002) estimated a possible reduction of flooding damage of about €54 billion through the development of emergency flooding areas at a cost of around €1 billion. In the same report, the total investment costs of installing all potential emergency storage areas is estimated at between €1.7 and 2.3 billion. This would provide a storage capacity of 1–1.2 billion m³. Actual flooding of these areas would, however, result in collateral damage of a maximum of €7.4 billion. It should be noted, however, that the social aspects are by far the most important factor in implementing these measures, as it would involve huge efforts from the regional population (capital losses, movement of companies) to develop these areas.

Conclusions

The impacts of the gradual shifts in temperature and precipitation due to climate change in the Rhine Basin seem limited in magnitude, especially in the near future. The increased climate variability on the other hand has more immediate and direct economic and safety (floods/droughts) implications.

In view of the magnitude of the perceived 'climate change threat', policy makers increasingly see that a continuation of the historical 'struggle against water' no longer involves pure technical and structural measures, but requires a different attitude in spatial planning and awareness, where space will be reserved to 'live with water'. However, in the translation of principles to practice, the recognition of the real socio-economic consequences of the new approach, in combination with a reduced feeling of urgency, is likely to result in a continuation of traditional measures based on short-term interventions and investments. For the Rhine Basin this may mean that strengthening of dykes and levees will still be considered preferable to providing 'space for the river'. The main driving force is socio-economic. Climate change will act as a moderating factor, but as long as the impacts of climate change and climate variability are mainly at the level of 'nuisance factors', acceptance and implementation of adaptation measures will be insignificant.

In this regard an industrialized basin like the Rhine has its limits; due to the relative safety against floods after centuries of protective measures, it is difficult to abandon existing approaches in favour of compromises with major consequences and uncertain benefits. The need for protection against extremes (and the willingness to face the consequences) in fact only becomes a reality after a (near) disaster. In those cases, however, the push for short-term emergency measures will prevail over the long-term approaches that would be more sustainable.

Apart from the conflict between traditional approaches and adaptation measures, the cost–benefit aspect will be the most important factor, with emphasis on the financial component but implicitly or explicitly including social costs (damage to goods, loss of life, intangible costs, etc.). The economic position of a country or region and the associated political priorities will be determining factors for the feasibility of activities. New and long-term initiatives are more likely to be accepted during periods of economic growth, with the traditional approaches preferred during periods of economic hardship.

Specific conclusions

- The expected impacts of climate change in the Rhine Basin are relatively modest. With regard to agriculture, some regions may suffer from increased water stress, but in general there actually may be an increase in productivity. For the environment, species in marginal situations may either benefit (increase in Mediterranean species) or disappear (southerly locations of boreal species). Increased risk to humans is especially related to the increased climate variability, although the ecological changes may include spreading of vector diseases. For industry, limited availability of cooling water during drought periods and reduced navigability during periods of low flow are the most likely concerns.
- In the Rhine Basin, the EU policies, especially concerning agriculture, are more significant than the effects of climate change.
- Increased climate variability, with a higher frequency of extreme events, is a more significant factor in decision making than the long-term gradual effects of climate change.
- Climate change and the associated increased climate variability are just one of the driving forces in the definition and especially the implementation of coping and adaptation measures. Socio-economic considerations and conditions are the dominant factor.
- In view of the innovative nature of adaptation measures and strategies, these will be most successfully developed and initiated during periods of economic growth. At other times more traditional approaches are likely to prevail, as they are considered cheaper, of proven reliability and a low risk investment.
- In the Rhine Basin, relatively high protection against floods along the river and in its estuary is secured by technological measures. This security is, compared to other basins, a limiting and inhibiting factor for innovative approaches in the management of the river basin, even though it is based on a false sense of security. In the Meuse Basin there was considerable disbelief when an approximately once in 150 years flood occurred twice within a few years. In this case the second flood, however, only caused a fraction of the damage (and the public agitation) of the first one.

References

CHR (1993) *Der Rhein unter der Einwirkung des Menschen; Ausbau, Schiffahrt, Wasserwirtshaft.* Report I-11. International Commission for the Hydrology of the Rhine, Lelystad, The Netherlands.

Commissie Noodoverloopgebieden (2002) *Eindrapport.* Projectsecretariaat Commissie Noodoverloopgebieden, The Hague.

Dieperink, C. (1997) International regime development: lessons from the Rhine catchment area. *Development Research Institute Quarterly Review, Thailand* 12, 27–37.

European Commission (2003) *Reform of the Common Agricultural Policy. A Long-term Perspective for Sustainable Agriculture, Impact Analysis.* European Commission, Directorate-General for Agriculture, Brussels.

Grabs, W. (1997) *Impact of Climate Change on*

Hydrological Regimes and Water Resources Management in the Rhine Basin. Report I-16. International Commission for the Hydrology of the Rhine, Lelystad, The Netherlands.

ICPR (2002) *Non Structural Flood Plain Management; Measures and their Effectiveness.* International Commission for the Protection of the Rhine, Koblenz.

Kwadijk, J.C.J. (1993) The impact of climate change on the discharge of the River Rhine. PhD thesis, Universiteit Utrecht, vakgroep Fysische Geografie. KNAG/NGS publicatie 171.

Middelkoop, H., Asselman, N.E.M., Buitenveld, H., Haasnoot, M., Kwaad, F.J.P.M., Kwadijk, J.C.J., van Deursen, W.P.A., van Dijk, P.M., Vermulst, J.A.P.H. and Wesseling, C. (2000) *The Impact of Climate Change on the River Rhine and the Implications for Water Management in the Netherlands.* Report 410200049. RIVM, De Bilt.

Middelkoop, H., van Asselt, M.B.A., van't Klooster, S.A., van Deursen, W.P.A., Haasnoot, M., Kwadijk, J.C.J., Buiteveld, H., Können, G.P., Rotmans, J., van Gemert, N. and Valkering, P. (2001) *Development of Flood Management Strategies for the Rhine and Meuse Basins in the Context of Integrated River Management.* Report of the IRMA-SPONGE project 3/NL/1/164 / 99 15 183 01. ICIS-Maastricht University/Dept of Physical Geography, Utrecht University.

NWP (2003) *Climate Adaptation in Water Management – How Are the Netherlands Dealing With It?* Discussion report on the Dutch Dialogue on Climate and Water. Hooijer, A., Kerssens, P., Balfoort, H., Klein, H., Kattenberg, A. and de Boer, M. (eds) NWP, Delft.

RIZA (2000) Visions for the Rhine (draft). Regional vision document for the World Water Council. Institute for Inland Water Management and Waste Water Treatment. RIZA, Lelystad, The Netherlands.

Speafico, M. and Kienholz, H. (1996) *The Rhine River: Lifeline of Large Cities in the Basin: a Case Study.* World Meteorological Institute, Geneva.

Stephens, W., Hess, T. and Knox, J. (2001) *Review of the Effects of Energy Crops on Hydrology.* Institute of Water and Environment, Cranfield University, Silsoe, UK.

WB21 (2000) *Water beleid voor de 21e eeuw. Advies van de Commissie Water beheer 21e eeuw.* Verkeer en Waterstaat, The Hague.

8 Will We Produce Sufficient Food under Climate Change? Mekong Basin (South-east Asia)

CHU THAI HOANH,[1] HANS GUTTMAN,[2] PETER DROOGERS[3] AND JEROEN AERTS[4]

[1]International Water Management Institute (IWMI), Colombo, Sri Lanka; [2]Mekong River Commission Secretariat, Phnom Penh, Cambodia; [3]FutureWater, Arnhem, The Netherlands; [4]Institute for Environmental Studies, Vrije Universiteit Amsterdam, Amsterdam, The Netherlands

Introduction

Critical issues

The Mekong is one of the world's largest river basins (IMC, 1988) with an area just under 800,000 km². It is shared by six countries: China (Yunnan Province), Union of Myanmar, Lao PDR, Kingdom of Thailand, Kingdom of Cambodia and Vietnam (Fig. 8.1). The basin is the home to over 65 million inhabitants (MRC and UNEP, 1997). The Lower Mekong River Basin (LMB) covers 77% of the Mekong River Basin (MRB) and is regarded as the most important part of the Mekong Basin, both environmentally and economically. The population in LMB is largely rural and most people are employed in agricultural or related activities, with rice as the major crop. Agricultural production is one of the five key areas of natural resources and development in the region (Nilsson and Segnestam, 2001). The other areas are hydropower generation, fisheries, forest resource management and the use of biological resources for conservation, tourism, trade and local livelihoods.

Although the Mekong riparian countries enjoy abundant water resources, availability varies widely across regions and seasons due to the monsoon rainfall pattern. Competition for scarce water resources is particularly evident during the dry season. There are several emerging issues regarding water usage in the MRB: (i) irrigation development in North-east Thailand has resulted in a lack of water during the dry season and restricts the amount of dry season cropping possibilities; (ii) intrusion of saline water to the Mekong Delta in Vietnam depends on the magnitude of the dry-season flows from upstream and the level of abstractions for irrigation in the delta; (iii) the floods; (iv) hydropower development, built or proposed, with distribution of

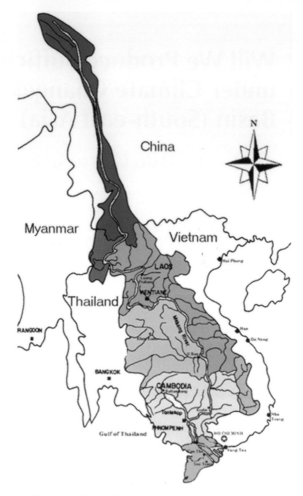

Fig. 8.1. Location of Mekong River Basin.

benefits across countries and sectors not thoroughly evaluated; and finally, (v) growing populations and increasing economic development is starting to affect water availability negatively by degrading the water quality (MRC, 2001).

    With a high population density per rice land (4.7 persons/ha in the LMB) and severe floods during the last few years, in this study we focus on two issues in the MRB: food production and flood under climate change (CC).

Physical characteristics

The MRB is located in South-east Asia where the climate is governed by monsoons – steady winds that blow alternately from the northeast and the southwest, each for about half of the year. The southwest monsoon begins in May and continues until

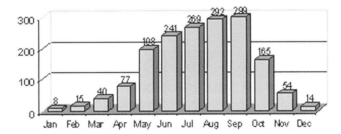

Fig. 8.2. Mean rainfall distribution (mm) in the LMB.

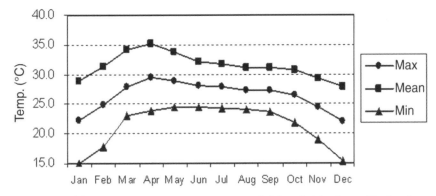

Fig. 8.3. Monthly temperature in 1961–1990 at Mukdahan in North-east Thailand.

late September. The northeast monsoon is from November to March. April and October are transitional periods with unstable wind speed and direction.

The regional rainfall varies significantly from the driest region in the basin (North-east Thailand), where annual rainfall is mostly between 1000 and 1600 mm, to the wettest regions (northern and eastern Highlands) with 2000–3000 mm annually. During the southwest monsoon (May to October), corresponding to the rainy season, the basin receives most of its annual rainfall, although it is somewhat shielded by the coastal mountains of Thailand and Cambodia (Fig. 8.2). Cyclonic disturbances may cause widespread rainfall of long duration during July to September, which can cause serious flooding.

Throughout the LMB, the air temperature is remarkably uniform, with small variations due to elevation and seasonal and maritime influences (IMC, 1988). Temperatures in the LMB are high, except during the early part of the northeast monsoon (November–December) when the winds from Central Asia bring somewhat cooler air. Then, temperature gradually rises up until February when, under the influence of light southerly winds, the weather becomes very hot. This high temperature lasts until the south-west monsoon commences in May (Fig. 8.3).

The MRB comprises of seven principal landform sections with different characteristics. The Mekong River has abundant surface water resources, with a total of approximately 475,000 million m³/year (Table 8.1). The catchment of Lao PDR contributes 35% of this total; Yunnan, Thailand and Cambodia catchments each

Table 8.1. Distribution of rainfall and runoff in MRB catchment by country. Source: ENSIC (1999).

Description	Catchment inside MRB						MRB[a] Total
	Yunnan	Myanmar	Lao PDR	Thailand	Cambodia	Vietnam	
Catchment area (km^2)	147,000	24,000	202,000	184,000	155,000	65,000	777,000
Catchment area as % of total MRB[a]	22	3	25	23	19	8	100
Average rainfall (mm/year)[b]	1,561	–	2,400	1,400	1,600	1,500	1,750[c]
Average flow (m^3/s) from area	2,414	300	5,270	2,560	2,860	1,660	15,060
Average runoff (million m^3)	76,128	9,461	166,195	80,732	90,193	52,350	474,932
Dry season runoff (million m^3)	19,032	1,419	24,929	12,110	13,529	7,852	71,240
Average runoff as % of total MRB	16	2	35	17	19	11	100

[a] MRB excluding the part inside Qinghai province and the Tibet Autonomous Region of China.
[b] Value by country estimated by different authors (see ENSIC, 1999).
[c] The annual total by IMC (1988) is only 1672 mm/year.

contribute between 15% and 20%; the Vietnamese catchment contributes just over 10% and the lowest contribution is from Myanmar, with only 2%. The highest runoff is observed in the Lao PDR west of the Anamite Mountains.

Socio-economic and institutional characteristics

The most notable economic feature of countries in the LMB is the difference in average levels of GNP between the wealthiest and poorest countries (Ringler, 2001). A key issue in Mekong Basin development and management is the cross-sectoral implications of interventions (Hirsch and Cheong, 1996). The population of the MRB is mainly agrarian-based, with rice as the major crop. The proportion of arable land under irrigation in the LMB is fairly low: about 20% in Laos, Thailand and Cambodia and 50% in the Vietnamese Mekong Delta, suggesting good opportunities for expansion. Vietnam on the whole retains only some 20% of its original moist forests compared to 43% for Thailand, 55% for Lao PDR and 71% for Cambodia. The LMB is considered to support one of the richest river faunas in the world. Fish production in the LMB is predominantly from capture fisheries (about 90%) and total production is estimated to range from 624,301 to 887,473 tonnes. However, including small-scale fisheries, it was estimated that actual catches may exceed 1 million tonnes per year (Hirsch and Cheong, 1996). Fish is the primary source of animal protein in the LMB and comprises from 40 to 80% of the total animal protein intake (IMC, 1988, 1992).

As early as 1947, preliminary studies and reports recognized the need for an international organization to coordinate activities and promote cooperation among the nations of the MRB. The Committee for Coordination of Investigations of the Lower Mekong Basin was formed in 1957 with four member countries: Lao PDR, Thailand, Cambodia and Vietnam, with a general, but limited, mandate to 'promote, coordinate, supervise and control the planning and investigations of water resource development projects' (Radosevich and Olson, 1999). In the decades that followed, the Mekong Committee focused on both short-term tributary projects and longer-term mainstream projects, but the development of the tributaries was regarded as the most important short-term issue. In 1978 the Mekong Committee was replaced by the Interim Committee for Coordination of Investigations of the Lower Mekong Basin with three members: Lao PDR, Thailand and Vietnam (MRC Secretariat, 1989), and in 1991 Cambodia requested re-admission and reactivation of the Mekong Committee. Then, the Agreement on the Cooperation for the Sustainable Development of the MRB (Mekong Agreement) was signed in 1995. The Mekong River Commission (MRC) replaced the Interim Mekong Committee with a new structure and with an agenda to address a host of environmental, social and economic challenges, and is expected to take on a strong role in coordinating planning, development and management of the MRB (MRC, 1995).

Projection of population and food demand

In this study, we applied two different population growth rates for scenarios A2 and B2 to project the population in 2025 and 2085, the middle years of two periods 2010–2039 and 2070–2099, with the assumption that population in A2 will be higher than in B2 (Table 8.2). Based on population in 2000, population in A2 in 2025 and 2085 are 93.0 and 186.2 million, respectively, while in B2 these numbers are 86.5 and 137.8 million in 2025 and 2085. Food demand is assumed to be increased in proportion with population growth. With the assumption of unchanged individual food demand of 300 kg per capita/year (ENSIC, 1999), overall food demand will be increased by 1.3–1.4 times in 2025 and 2.1–2.9 times in 2085 (Table 8.2).

Climate Change Scenarios

In the ADAPT project, climate change projections for the periods 2010–2039 and 2070–2099 under two IPCC scenarios are selected for comparison with the baseline period of 1961–1990. From the many global models producing climate change projection data, we selected to use the SRES A2 and B2 projections by the model from the Hadley Centre for Climate Prediction and Research in the UK, referred to as HadCM3. For the baseline period 1961–1990, temperature and precipitation data from the Climatic Research Unit were used. For more information see Chapter 2.

Comparison of climate data during 1961–1990 in different data sets

For the comparison between current climate and projected climate, we looked at the differences in average temperature and precipitation at both basin level and regional

Table 8.3. Comparison of temperature and precipitation data from CRU and SRES data sets.

Item	CRU 1961–1990	SRES 1961–1990	A2 2010–2039	A2 2070–2099	B2 2010–2039	B2 2070–2099
Mean temperature	25.3	24.3	25.3	28.3	25.3	27.2
SD (°C)	0.3	0.4	0.5	0.9	0.5	0.6
CV (%)	1.1	1.8	2.0	3.0	2.0	2.1
Mean precipitation (mm/year)	1713	1563	1534	1623	1530	1546
SD (mm)	142	135	156	191	145	180
CV (%)	8.3	8.6	10.2	11.8	9.5	11.6

maximum monthly temperature will be over 40°C and the minimum temperature will be over 25°C, and the number of these months is higher in A2 than in B2.

During 2010–2039 in scenario A2, mean temperature in every sub-basin will increase about 3.7–4% compared with the baseline 1961–1990. The same increase is found during 2010–2039 in scenario B2. During 2070–2099, the increments in mean temperature in scenario A2 by sub-basin vary from 14.0% (Delta) to 21.8% (Nam Ou) compared with the baseline 1961–1990. However, in this period, the increments in mean temperature in scenario B2 are lower, from 10.0% (Delta) to 15.7% (Nam Ou).

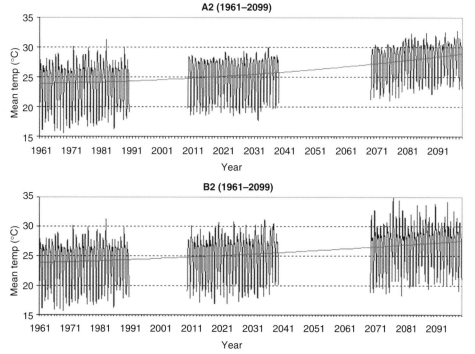

Fig. 8.5. Variations and trend of mean monthly temperature of the MRB under scenarios A2 and B2.

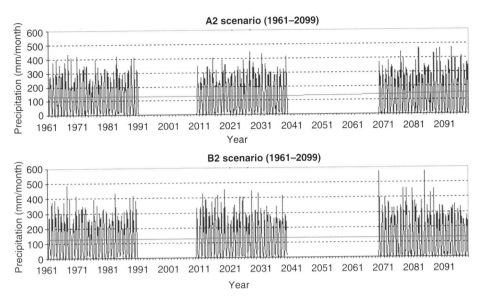

Fig. 8.6. Variations and trend of monthly precipitation of the MRB under scenarios A2 and B2.

Precipitation

Changes of mean precipitation from 1961–1990 to 2010–2039 and 2070–2099 in Mekong River sub-basins under scenarios A2 and B2 are shown in Fig. 8.6. Overall precipitation in the basin increases, although the change in precipitation varies from one sub-basin to another. Compared with the baseline 1961–1990, during 2010–2039, mean precipitation (from here onwards, the mean precipitation refers to the value after adjustment) in different sub-basins varies from about −6% to +6% in both scenarios A2 and B2. However, for the MRB as a whole, mean precipitation during 2010–39 only varies by +0.2% and −0.2% in scenarios A2 and B2, respectively. During 2070–2099, mean precipitation in different sub-basins varies from about −12% to +32% in both scenarios A2 and B2. The positive variations occur in nine sub-basins with a higher percentage compared to negative variations in four sub-basins. These variations lead to an increase of 9.8% and 9.4% in the MRB as a whole in scenarios A2 and B2, respectively.

Regarding other factors that affect crop production and water balance, there are three different trends from 1961–1990 to 2010–2039 and 2070–2099: (i) minor change in solar radiation; (ii) increase in wind speed; and (iii) decrease in relative humidity. In all cases, the variations in A2 are larger than in B2.

Impacts of Climate Change

Modelling activities

In the ADAPT project, the SWAP 2.0 model is applied to analyse the variation of food production at field-scale level. SWAP is a one-dimensional physically based model for water, heat and solute transport in the saturated and unsaturated zones, and also includes modules for simulating irrigation practices and crop growth (see Chapter 3). The SLURP model has been used for hydrological modelling on the basin scale (Kite, 2000). SLURP is a hydrological basin model that simulates the hydrological cycle from precipitation to runoff, including the effects of reservoirs, regulators, water extractions and irrigation schemes. It first divides a basin into sub-basins using topography from a digital elevation map.

In this study, we used the previously described data from climate change scenarios comprising three data sets for 1961–1990, 2010–2039 and 2070–2099. However, since the SLURP model requires daily climate data, the SRES scenarios (provided at a monthly time step) were interpolated to daily scenario values. The process was applied for two main parameters (temperature and precipitation). To analyse the effects of climate change, values of other factors such as land cover area, reservoir, etc. are kept the same as in the previous SLURP/MRB study (Kite, 2000).

Impacts of climate change

Projected climate changes in the MRB include strengthening of monsoon circulation, increases in temperature and increases in the magnitude and frequency of extreme rainfall events (IPCC, 2001). Climate-related effects also include sea-level rise. These changes could result in major impacts on the region's ecosystems and biodiversity; hydrology and water resources; agriculture, forestry and fisheries; mountains and coastal lands; and human settlements and human health. The following are major features of climate changes that affect the MRB.

- The monsoons bring most of the region's precipitation and are the most critical climatic factor in the provision of drinking water and water for rain-fed and irrigated agriculture. As a result of the seasonal shifts in weather, a large part of Tropical Asia, including the MRB, is exposed to annual floods and droughts.
- Tropical cyclones, frequency and related floods may increase.
- Other extreme events include high-temperature winds.
- In the megacities and large urban areas, high temperatures and heat waves also occur. These phenomena are exacerbated by the urban heat-island effect and air pollution.
- The El Niño-Southern Oscillation (ENSO) phenomenon becomes geographically much more extensive, which has an especially important influence on the weather and interannual variability of climate and sea level, especially in the western Pacific Ocean and South China Sea around the MRB.

The main current water issues in the MRB may be accelerated through climate change (WUP, 2001).

- Water shortages in Thailand: the level of irrigation development in the Northeast Thailand region, as well as in neighbouring, agriculturally important Chao Phraya Basin in Thailand, has resulted in a lack of water during the dry season.

- Salinity intrusion in delta: the extent of the intrusion of saline water into the Mekong Delta depends on the magnitude of the dry-season flows from upstream and the level of abstractions for irrigation.
- Floods: the flood in 2000 was the highest flood in over 40 years, with significant loss of life and high damage to crops and infrastructure in Cambodia and Vietnam.

Impacts of climate change on food production

Changes in climate will have significant effects on agriculture in many parts of the MRB, particularly on low-income populations that depend on isolated agricultural systems. These effects are due to frequent floods, droughts, cyclones, sea-level rise, higher temperature that can damage life and property and severely reduce agricultural production. Two additional factors increase the vulnerability of the agricultural sector in the LMB under climate change: (i) only about 7–10% of the cultivated land in the LMB is irrigated (Ringler, 2001). Yields for rice under rain-fed conditions in the LMB are low, compared with those of irrigated rice yields, due to many factors, including floods, droughts, temporary inundation from rainfalls, and tidal flows and coastal salinity (Hossain and Fischer, 1995); (ii) the region supports a large human population per hectare of cropland, from 2.4 in North-east Thailand to 11.5 in the Vietnamese Mekong Delta.

Impacts of climate change on the environment

As in other regions in Tropical Asia, climate change represents an important additional external stress on the numerous ecological and socio-economic systems in the MRB that already are adversely affected by air, water or land pollution, as well as increasing resource demands, environmental degradation and unsustainable management practices (IPCC, 2001). The main current environmental issues related to climate change in the MRB are closely linked to water usage, particularly regarding availability of water for agriculture, but also to flooding, and storage and release of water from hydropower dams.

Indicators

Knowing the major hydrological, food and environmental issues related to climate change, we now can identify a set of indicators that allows for a quantitative impact assessment exercise. In this study we focus on a set of indicators for the general basin hydrology. An increase in *maximum flows*, in particular *daily flow*, means a higher risk of flood, while a decrease in *minimum* or *average monthly flow* means a risk of water shortage for certain land use types and deeper salt intrusion into the Delta. Furthermore, we used two indicators that measure the state of food security in the MRB and two indicators that measure the quality of the environment. The selected indicators are listed in Table 8.4. It is noted that this list is not final and many more important indicators could be considered in subsequent studies.

Table 8.4. List of selected indicators for the MRB.

Group	Indicator	Description
Hydrological indicators	Maximum daily flow (m³/s)	Output from the SLURP model
	Maximum, minimum and average monthly flows (m³/s)	Outputs from the SLURP model
Indicators for food security	Total rice production (t/year)	Estimated from current production with the assumption that it will alter in proportion with changes in yield predicted by SWAP model and crop intensification
	Farm income index (relative to 1)	Equal to 1 for 1961–1990 and will alter in proportion to yield increased by climate changes and crop intensification
Indicators for environment	Number of people affected by floods	Estimated from current people affected in each country with the assumption that it will alter in proportion to the increment in maximum daily flow
	Salinity intrusion (+++/– – –)	Qualitatively evaluated based on minimum monthly flow into the delta

Adaptation Strategies

Possible adaptation measures and expected qualitative effects on food production and environment in the MRB are shown in Table 8.5. There are many measures that have positive and significant effects on either food production or environment, or both, but several effects are unknown, in particular on the environment.

Two major adaptation strategies are analysed for the MRB. They will be compared to a Business As Usual (BAU) scenario where we expect no major changes in climate conditions. However, under BAU, food production will be further developed to meet the increased demand due to population growth. Also, a scenario with climate change (CC) will be analysed, but still without taking specific adaptations. Table 8.6 provides an overview of all scenarios and strategies.

Adaptation strategy 1 'CC-Agri' is developed to enhance food security under climate change. Specific measures under this strategy are to intensify crops, because most land suitable for agriculture has been exploited. The agricultural area in the sub-basins is unchanged under this strategy. It is expected that food demand will increase proportionally with population number during 2010–2039 and 2070–2099. Values in the middle years, 2025 and 2085, represent the situation in each period. Another measure is to expand the current crop calendar from 1 June/30 September to 1 June/31 December in 2010–2039 and 1 June/31 January in 2070–2099. More irrigation water is needed for such expansion of crop season, i.e. the stream flow in the low flow season will be affected.

Adaptation strategy 2 'CC-Agri-For' focuses on enhancing environmental quality under CC and proposes to implement large-scale reforestation activities to alleviate impacts from floods and soil erosion. The shrub land (17.7% of basin area) is assumed to be fully covered by mixed forest.

Table 8.5. Possible adaptation measures for food production and environmental protection in the MRB.

	Expected effects on	
Adaptation measures	Food production	Environment
Improve water retention capacity of agricultural land and forest watersheds	P, S	P, I
Improve farming and irrigation techniques at field scale	P, S	U, I
Improve irrigation management at sub-basin scale	P, S	P, S
Use storage capacity of natural water bodies	P, S	P, S
Increase irrigation development and focusing on irrigation in multi-purpose projects	P, S	U, I
Improve livelihood of food-producing farmers by socio-economic interventions such as price management, extension services, marketing, etc.	P, S	U, I
Integration between water resource development in LMB with that in Upper MRB	P, S	P, S
Reforestation on shrub land	P, I	P, S
Manage salinity intrusion in the delta	P, S	P, S
Preserve the productivity of the aquatic ecosystem	U, I	P, S
Diversify cropping patterns	N, S	P, S
Manage and mitigate damage from floods	P, I	P, S
Increase hydropower development	P, I	N, S
Increase usage of navigation for trade	P, I	U, I

P, positive; N, negative; U, unknown; S, significant; I, insignificant.

Evaluation of Impacts and Adaptation

The analysis focuses on two sub-basins: (i) the Mekong 3, which receives water from the entire upstream part of the MRB; and (ii) the Delta, which collects water from the whole basin, including the Tonle Sap Lake in Cambodia.

Business as usual (No CC, No adaptation)

Food

In this scenario, food demand in the basin will increase in proportion with population growth as in Table 8.3. Crop intensification is applied to satisfy future food demand as described above.

Impacts on water resources (Table 8.7)

In this scenario, maximum daily flow as well as maximum monthly flow does not change much (about 1%) compared to 1961–1990. However, the expansion of crop season into

Table 8.6. Scenarios for climate change and adaptation strategies.

No.	Scenario	Adaptation strategy	Objective	Agricultural adaptation measures	Environmental adaptation measures
1	BAU	No adaptation	Current trend without CC	Crop intensification is needed to satisfy food demand for growing population	No adaptation
2	2010–2039 for A2 and B2 2070–2099 for A2 and B2	No adaptation	Effects of CC	No change in agriculture	No adaptation
3	2010–2039 for A2 and B2 2070–2099 for for A2 and B2	CC-Agri	Effects of CC and enhancing food security	Crop intensification is needed to satisfy food demand growing population	No adaptation
4	2010–2039 for A2 and B2 2070–2099 for A2 and B2	CC-Agri-For	Effects of CC and enhancing environmental quality	Crop intensification is needed to satisfy food demand for growing population	Reforestation of all shrub land to mixed forest

Table 8.7. Impacts on water resources in scenario Baseline (BAU) for sub-basins Mekong 3 and Delta.

Scenario / sub-basin		Indicator	1961–1990 m³/s	2010–2039 m³/s	2010–2039 % change	2070–2099 m³/s	2070–2099 % change
BAU/Mekong 3	Hydrology	Min. monthly flow	560	463	−17	0	−100
		Max. monthly flow	40,995	41,214	+1	41,031	+0
		Ave monthly flow	12,894	11,976	−7	11,694	−9
BAU/Delta	Hydrology	Min. monthly flow	1,857	1,588	−15	151	−92
		Max. monthly flow	45,681	46,047	+1	45,861	+0
		Ave monthly flow	13,495	13,777	−11	13,193	−15
BAU/Mekong 3	Hydrology	Max. daily flow	54,829	55,111	+1	54,932	+0
BAU/Delta	Hydrology	Max. daily flow	51,161	51,559	+1	51,359	+0

Underlined values in 2010–2039 and 2070–2099 are higher than values in baseline 1961–1990. % change is compared to 1961–1990.

low flow season has significant impacts on the minimum flow by a reduction of about 15–17% in 2010–2039 and 90–100% in 2070–2099. This reduction will cause serious water shortage at upstream sub-basins and salinity intrusion in the Delta. Therefore measures at a different scale will be needed, such as water-saving irrigation techniques at field scale or reserving more water for low-flow season at sub-basin level. At present, the MRCS is beginning a study on integrated basin flow management (MRC, 2003b) with the objectives indicated in the 1995 Agreement: maintenance of flows on the mainstream of the Mekong River: (i) of not less than the acceptable minimum monthly natural flow during each month of the dry season; (ii) to enable the acceptable natural reverse flow of the Tonle Sap during the wet season; and (iii) to prevent average daily peak flows greater than those that naturally occur on average during the flood season. Extracting more water for irrigation in the dry season may cause a decrease of about 10–15% in the average stream flows in 2010–2039 and 2070–2099.

Industry

A decrease in the average stream flows in 2010–2039 and 2070–2099 implies less water for hydropower. However, the reduction is only about 10–15% in these periods.

Effect of climate change without adaptation

Although this scenario will not happen in reality without making suitable adaptations, it is calculated in order to analyse the effect of CC alone on the hydrology of the MRB, with the assumption that all other factors such as land use are the same as at present.

Impacts on water resources

The trend and monthly variations in stream flow of sub-basin Mekong 3 due to climate change are shown in Fig. 8.7. Although the general trend shows only slight increases, higher extremes are expected for the future. In A2 (Table 8.8), maximum monthly flows in some sub-basins increase in 2010–2039 compared to the 1961–1990 period (14% in Mekong 3) and the Baseline scenario. In 2070–2099 the increments are even higher (41% in Mekong 3 and 19% in the Delta). In B2 the increments are smaller than in A2 (10% and 35% in Mekong 3 in 2010–2039 and 2070–2099, respectively). The monthly average flow is almost unchanged in 2010–2039, but increases in 2070–2099 with a lower level than the maximum (17% in Mekong 3). Minimum monthly flow will decrease slightly in 2010–2039 (7% in A2 and 15% in B2 in the Delta), but significantly in 2070–2099 (26% in A2 and 29% in B2 in the Delta). In this scenario, maximum daily flow also increased significantly in both Mekong 3 and the Delta in 2070–2099, about 30% in A2 and 15% in B2.

Food

Under both the A2 and B2 scenarios, a positive impact is the increase of average flow, i.e. more water will be available for agriculture. Moreover, outputs from the SWAP model at the field scale show an increase in rice yield of 10% (2010–2039) to 40% (2070–2099) in A2, but less in B2, at 10% to 26%, respectively.

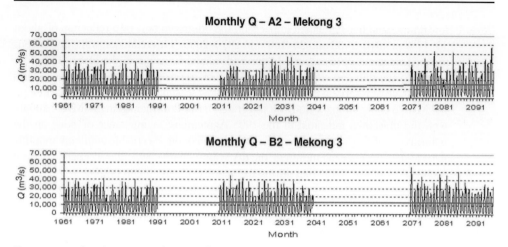

Fig. 8.7. Trend and variation of stream flow in sub-basin Mekong 3 under A2 and B2.

Environment

There are two negative impacts on the environment. An increase of maximum monthly flow, and in particular daily flow, means that there will be more severe floods in the MRB due to climate changes. On the other hand, a decrease in minimum monthly flow implies deeper salinity intrusion in the low flow season under the climate change scenarios. Compared to B2, the flood level is slightly higher in A2 and the salinity intrusion level is slightly less than in B2.

Industry

An increase in average monthly flow implies positive impacts on hydropower.

Adaptation strategy for enhancing food security (CC-Agri)

Impacts on water resources

The combined impacts of climate changes and agricultural development for food demand in the future are shown in Table 8.9. Maximum monthly and daily flows are not much different than in the scenario CC-NoAgri because only in the dry season more irrigation water is extracted. Reduction of monthly minimum flows will be worse, in particular for the B2 scenario, because both more water will be extracted for irrigation purposes and climate change will lower extreme minimum flows. On the other hand, the reduction in average stream flows due to irrigation is compensated by the increase due to climate change, therefore the variations in both A2 and B2 are only 5–10% compared to 10–15% in the Baseline scenario.

Food

Compared to the Baseline scenario, positive impacts of climate change on food production are the increase in rice yield and the increase of average flow that makes more water available for agriculture. However, the decrease in minimum flow will be a constraint to the expansion of the crop season into low flow seasons. Adaptation

Table 8.8. Impacts on water resources in scenario CC-NoAgri.

Scenario / sub-basin	Indicator	1961–1990 m³/s	2010–2039 m³/s	2010–2039 % change	2070–2099 m³/s	2070–2099 % change
A2/Mekong 3	Hydrology	Min. monthly flow	560	574	+2	424	−24
		Max. monthly flow	40,995	46,635	+14	57,659	+41
		Ave monthly flow	12,894	12,940	+0	15,147	+17
A2/Delta	Hydrology	Min. monthly flow	1,857	1,726	−7	1,370	−26
		Max. monthly flow	45,681	45,091	−1	54,269	+19
		Ave monthly flow	15,459	15,399	−0	17,698	+14
A2/Mekong 3	Hydrology	Max. daily flow	54,829	57,387	+5	74,055	+35
A2/Delta	Hydrology	Max. daily flow	51,161	50,294	−2	66,439	+30
B2/Mekong 3	Hydrology	Min. monthly flow	560	592	+6	466	−17
		Max. monthly flow	40,995	45,163	+10	55,168	+35
		Ave monthly flow	12,894	12,973	+1	13,672	+6
B2/Delta	Hydrology	Min. monthly flow	1,857	1,588	−15	1,311	−29
		Max. monthly flow	45,681	45,271	−1	53,181	+16
		Ave monthly flow	15,459	15,397	−0	16,140	+4
B2/Mekong 3	Hydrology	Max. daily flow	54,829	54,016	−1	62,118	+13
B2/Delta	Hydrology	Max. daily flow	51,161	50,188	−2	58,696	+15

Underlined values in 2010–2039 and 2070–2099 are higher than values in baseline 1961–1990. % change is compared to 1961–1990.

measures will be needed such as water-saving irrigation techniques or sub-basin flow management, as mentioned in the Baseline scenario. Furthermore, in this phase of the ADAPT study, impacts of climate change on yield of food crops are analysed based on the results of the SWAP field-scale model. For the Mekong Basin, two crops have been selected for analysis: rice and maize. Thailand and Vietnam, two of the riparian countries, are rice exporters number one and two, respectively, in the world market. Therefore, rice yields under different scenarios from the SWAP model are analysed in a location in the Mekong River Delta to reflect the impact of climate change on food production. For the A2 scenario, rice production will increase substantially in the future, about 10% in 2010–2039 to 40% in 2070–2099. The main reasons for the yield increase are changes in CO_2 concentration and temperature. The B2 scenario shows a small increase, about 10% in 2070–2099.

Environment

There are two negative impacts on the environment: floods and salinity intrusion. The flood impacts will be at similar levels to the scenario CC-NoAgri. An adaptation strategy is the reforestation of shrub land (17.7% basin area) by mixed forest as shown in scenario CC-Agri-For. The water shortage and salinity intrusion in the dry season will be more serious than in the Baseline scenario.

Industry

Climate change has a positive impact on hydropower due to the increase in average monthly flow that partly can compensate for the reduction due to irrigation.

Adaptation strategy for enhancing environmental quality (CC-Agri-For)

Impacts on water resources (Table 8.10)

Climate change may cause more severe flooding in the MRB. An adaptation strategy is to afforest the shrub land (17.7% of the basin area) by mixed forest. Compared with scenario CC-Agri, the effect on maximum monthly flow is significant in scenario A2, with about 10% reduction in Mekong 3 in 2070–2099, but less under scenario B2. The effect is not as strong as expected, possibly due to: (i) the high slope of the basin; and (ii) the large agricultural area. The effect on maximum daily flow is not as significant as that on maximum monthly flow. However, compared with scenario CC-Agri, reforestation strategy will cause an approximately 5–8% lower monthly average flow due to higher evapotranspiration from the forest areas.

Food

Compared with scenario CC-Agri, there are no significant effects on food production due to this reforestation strategy.

Environment

Maximum monthly and daily flows (except daily flow in B2 during 2070–2099) are lower than in scenario CC-Agri, i.e. less severe flooding in the basin. The effects in flood reduction are higher in CC scenario A2 than B2. However, the decrease in maximum flow is not high enough to reduce flooding significantly. It implies that more

Table 8.9. Impacts on water resources in scenario CC-Agri.

Scenario / sub-basin	Indicator		1961–1990 m³/s	2010–2039 m³/s	% change	2070–2099 m³/s	% change
A2/Mekong 3	Hydrology	Min. monthly flow	560	557	−1	31	−95
		Max. monthly flow	40,995	46,646	+14	57,723	+41
		Ave monthly flow	12,894	12,109	−6	13,977	+8
A2/Delta	Hydrology	Min. monthly flow	1,857	1,508	−19	346	−81
		Max. monthly flow	45,681	45,446	−1	54,495	+19
		Ave monthly flow	15,459	13,828	−11	15,447	−0
A2/Mekong 3	Hydrology	Max. daily flow	54,829	57,629	+5	74,178	+35
A2/Delta	Hydrology	Max. daily flow	51,161	50,533	−1	66,648	+30
B2/Mekong 3	Hydrology	Min. monthly flow	560	472	−16	0	−100
		Max. monthly flow	40,995	45,504	+11	55,106	+34
		Ave monthly flow	12,894	12,162	−6	12,560	−3
B2/Delta	Hydrology	Min. monthly flow	1,857	1,304	−30	27	−99
		Max. monthly flow	45,681	45,262	−1	53,138	+16
		Ave monthly flow	15,459	13,857	−10	13,947	−10
B2/Mekong 3	Hydrology	Max. daily flow	54,829	53,979	−2	62,070	+13
B2/Delta	Hydrology	Max. daily flow	51,161	50,165	−2	58,655	+15

Underlined values in 2010–2039 and 2070–2099 are higher than baseline values in 1961–1990. % change is compared to 1961–1990.

adaptation measures, such as protection for people, should be considered. On the other hand, minimum monthly flow also decreases, i.e. deeper salinity intrusion will occur.

Industry

Compared to scenario CC-Agri, a decrease in average monthly flow will cause lower potential for hydropower.

Evaluation of adaptations for both food and environment

The effects of climate change on indicators are shown in Table 8.11. Food production in the future will increase due to the increase in rice yield allowed by climate change and the agricultural development of crop intensification. Farm income will increase in proportion with the increase in rice yield by climate change. Reforestation as part of the environmental adaptation strategy reduces the number of people affected by flooding to less than 3%.

Although climate change may cause higher yields at the field level, to satisfy the food demand for the future population in both A2 with high population and B2 with lower population, crop intensification will be the trend of agricultural development in the MRB. At present, farmers have begun crop intensification in many parts of the MRB. In the Delta of Vietnam and Cambodia, three major cropping seasons for rice are currently applied: winter–spring or early season with full irrigation at the end of crop season (November to February), summer–autumn or midseason with supplementary irrigation at the beginning of crop season (June to September), and main rain-fed season (June to December), the long-duration wet season crop. Fifty-two per cent of the rice in the Mekong River Delta in Vietnam is grown in irrigated lowlands, with the remaining 48% grown under rain-fed conditions with supplementary irrigation (Maclean *et al.*, 2002). However, rice crop growing in the dry season in the MRB of Thailand is still limited.

Forestry management has become a highly controversial issue in the LMB because of the benefits from forests, the illegal exploitation and the environmental consequences of forest losses, including potential impacts on hydrological processes, water quality and soil erosion. The annual deforestation rate in the LMB is estimated to be 0.53%. With this rate, the LMB would have only about 20% of forest cover left at the end of 2100 (MRC, 2003a). Hence, stopping (illegal) deforestation in the MRB and afforesting instead is an important activity. Although reforestation may not contribute much to the reduction of flood flow as discussed above, other effects on environment such as reducing watershed degradation and maintaining biodiversity have been noticed. To implement this adaptation strategy, a reforestation plan is required with the support of government authorities from the national to the local level. Community management should be applied, i.e. the key roles should be given to those people who are living in the forest land.

Table 8.10. Impacts on water resources in scenario CC-Agri-For.

Scenario / sub-basin	Indicator		1961–1990 m³/s	2010–2039 m³/s	% change	2070–2099 m³/s	% change
A2/Mekong 3	Hydrology	Min. monthly flow	560	508	−9	26	−95
		Max. monthly flow	40,995	42,376	+3	53,370	+30
		Ave monthly flow	12,894	11,170	−13	12,814	−1
A2/Delta	Hydrology	Min. monthly flow	1,857	1,401	−25	346	−81
		Max. monthly flow	45,681	43,190	−5	52,768	+16
		Ave monthly flow	15,459	12,715	−18	14,109	−9
A2/Mekong 3	Hydrology	Max. daily flow	54,829	57,410	+5	71,874	+31
A2/Delta	Hydrology	Max. daily flow	51,161	49,928	−2	64,245	+26
B2/Mekong 3	Hydrology	Min. monthly flow	560	441	−21	0	−100
		Max. monthly flow	40,995	44,536	+9	54,967	+34
		Ave monthly flow	12,894	11,221	−13	11,596	−10
B2/Delta	Hydrology	Min. monthly flow	1,857	1,243	−33	10	−99
		Max. monthly flow	45,681	41,726	−9	52,735	15
		Ave monthly flow	15,459	12,743	−18	12,829	−17
B2/Mekong 3	Hydrology	Max. daily flow	54,829	53,068	−3	66,124	+21
B2/Delta	Hydrology	Max. daily flow	51,161	48,221	−6	59,099	+16

Underlined values in 2010–2039 and 2070–2099 are higher than baseline values in 1961–1990. % change is compared to 1961–1990.

Table 8.11. Indicators in scenario CC-Agri and CC-Agri-For.

				Future no adaptation (CC-Agri)		Environment adaptation (CC-Agri-For)	
Scenario		Indicator	Measured in	Current 2000	2010–2039	2070–2099	2010–2039	2070–2099
A2	Environment	No. people affected by flood	number	20.9	**30.6**	**66.7**	**29.1**	**64.0**
		Salinity intrusion	+++/– – –	– –	– – –	– – –	– – –	– – –
	Food	Tonnes of rice per year	number	19.4	*22.2*	*25.4*	*22.2*	*25.4*
		Average farm income	index	1.0	*1.2*	*1.4*	*1.2*	*1.4*
B2	Environment	No. people affected by flood	number	20.9	**28.8**	**48.9**	**27.1**	**46.3**
		Salinity intrusion	+++/– – –	– –	– – –	– – –	– – –	– – –
	Food	Million tonnes of rice per year	number	19.4	19.4	*21.4*	19.4	*21.4*
		Average farm income	index	1.0	1.0	*1.1*	1.0	*1.1*

Italic, good; shaded, decrease; regular, stable; bold, very bad.

Conclusions

The MRB is a large basin with specific physical environmental and complex socio-economic conditions. Therefore, the basin is vulnerable to changes in climate and other related environmental factors. The most significant effects of these changes are in water resources. Adaptation to these changes should be considered in an integrated assessment context. Such integration is needed because individual countries, regions, resources, sectors and systems will be affected by climate change not in isolation but in interaction with one another. By referring to country studies on specific issues the geographical integration can be realized through linkages between different functional regions and analysing the transboundary issues. This geographical integration is particularly important because six countries with different socio-economic conditions are sharing the Mekong water.

Fortunately, the riparian countries recognized the need for cooperation in natural resources management. The Mekong River Commission of LMB has a long history of over 50 years in the coordination of studying and planning for resource development. Recently, the concept of the Greater Mekong Sub-region also helps in bringing the two upper basin countries closer to the LMB countries in terms of sharing resources and cooperating in regional development.

From the comparison of observed data and data from the HadCM3, we conclude that adjustment is needed for a large basin like the MRB, in particular for precipitation. The observed data, therefore, become important in this type of study. Data from the HadCM3 show that there is significant change in temperature in the MRB, about 4% in 2010–2039 and 12% to 16% in 2070–2099 in different scenarios. During 2010–2039, precipitation is not much different to the baseline 1961–1990. However, during 2070–2099, precipitation can increase by up to 10%. Changes in temperature are more homogeneous throughout the whole MRB, while changes in precipitation are very different from one location to another.

In this study, modelling and analyses were done for the MRB to identify the adaptation measures for the current as well as possible future issues in food security and environmental management related to climate changes. Even higher crop yield is expected; crop intensification will be the trend of agricultural development in the MRB to satisfy the food demand of future population under both A2 and B2 scenarios (about 1.3–1.4 times the current level in 2025 and 2.1–2.9 times in 2085). At present, crop intensification has begun in many parts of the MRB.

Outputs from the SLURP model for the 'business as usual' scenario without climate change but with crop intensification show that the expansion of the crop calendar into the dry season will have significant impacts on the minimum flow by a reduction of about 15–17% in 2010–2039 and 90–100% in 2070–2099. This reduction will cause serious water shortage at upstream sub-basins and salinity intrusion in the Delta, therefore measures at different scales will be needed, such as applying water-saving irrigation techniques at a field scale or reserving more water for the low flow season at the sub-basin level. A future decrease in the average stream flows of 10–15% also implies that less water will be available for hydropower.

Under climate change scenarios with different adaptation strategies for food and the environment, an increase in average monthly flow is expected, in particular at the most downstream sub-basins. However, an increase of about 10% will not be enough to compensate the higher irrigation requirements for food in the future. On the other hand, climate change will increase the maximum monthly and daily flows that cause higher flood levels up to 30–40% in 2070–2099. Moreover, minimum monthly flow will decrease about 20–30% during that period, i.e. more serious water shortage at upstream sub-basins and deeper salt water intrusion in the Delta will occur during the low flow season. The adaptation strategy of reforestation on shrub land in the whole basin will not help much in reduction of severe flood under climate change (less than 10%), possibly due to the high slope of the basin and the large agricultural area in the basin. However, other effects of this strategy on the environment such as reducing watershed degradation and maintaining biodiversity have been noticed. To implement this adaptation strategy, a reforestation plan is required with the support from government authorities at both national and local levels, i.e. the key roles should be given to those people who are living in the forest land.

Further analysis will be needed to incorporate changes in land use at field and basin levels as well as incorporate socio-economic conditions into the modelling activities. Also more adaptation strategies such as changing cropping pattern and the irrigation calendar, or increasing the storage capacity, etc. in different sub-basins will need to be considered in detail to deal with the higher flood level in the rainy season and the lower flow in the dry season.

References

ENSIC (1999) *Directions to Sustainable Water Management: Mekong River Basin*. Environmental Systems Information Center (ENSIC), Asian Institute of Technology (AIT), Bangkok.

Hirsch, P. and Cheong, G. (1996) *Natural Resource Management in the Mekong River Basin: Perspectives for Australian Development Cooperation*. Final overview report to AusAID. University of Sydney, Australia.

Hossain, M. and Fischer, K.S. (1995) Rice research for food security and sustainable agricultural development in Asia: achievements and future challenges. *GeoJournal* 35, 286–298.

IMC (1988) *Perspectives for Mekong Development. Revised Indicative Plan (1987) for the Development of Land, Water and Related Resources of the Lower Mekong Basin*. Committee report, Interim Committee for Coordination of Investigations of the Lower Mekong Basin, Bangkok.

IMC (1992) *Fisheries in the Lower Mekong Basin – Main Report*. Interim Committee for Coordination of Investigations of the Lower Mekong Basin, Bangkok.

IPCC (2001) *Climate Change 2001 – Synthesis Report. An Assessment of the Intergovernmental Panel on Climate Change*. Watson, R.T. (ed.) Cambridge University Press, Cambridge.

Kite, G. (2000) *Developing a Hydrological Model for the Mekong Basin. Impacts of Basin Development on Fisheries Productivity*. Working Paper 2. International Water Management Institute (IWMI), Colombo, Sri Lanka.

Maclean, J.L., Dawe, D.C., Hardy, B. and Hettle, G.P. (2002) *Rice Almanac*. International Rice Research Institute (Philippines), West Africa Rice Development Association (Cote d'Ivoire), International Center for Tropical Agriculture (Colombia) and Food and Agriculture Organization (Italy).

MRC (1995) *Agreement on the Cooperation for the Sustainable Development of the Mekong River Basin*. April. Mekong River Commission (MRC), Phnom Penh.

MRC (2001) *MRC Hydropower Development Strategy. Meeting the Needs, Keeping the Balance*. Mekong River Commission (MRC), Phnom Penh.

MRC (2003a) *Status of the Basin*. Draft March. Mekong River Commission (MRC), Phnom Penh.

MRC (2003b) *Water Utilization Program Start-up Project: Integrated Basin Flow Management*. Draft 21 May. Mekong River Commission (MRC), Phnom Penh.

MRC Secretariat (1989) *The Mekong Committee: a Historical Account (1957–89)*. Interim Committee for Coordination of Investigations of the Lower Mekong Basin, Bangkok.

MRC and UNEP (1997) *Mekong River Basin Diagnostic Study – Final Report*. Mekong River Commission (MRC), Bangkok and United Nations Environment Programme (UNEP).

Nilsson, M. and Segnestam, L. (2001) *The Institutional Challenge for Natural Resource Use and Development in the Mekong Region*. SEI/REPSI Report Series No. 1. Stockholm Environment Institute, Sweden.

Radosevich, G.E. and Olson, D.C. (1999) Existing and emerging basin arrangements in Asia: Mekong River Commission case study. Third Workshop on River Basin Institution Development. The World Bank, Washington, DC.

Ringler, C. (2001) *Optimal Allocation and Use of Water Resources in the Mekong River Basin: Multi-country and Intersectoral Analyses*. Development Economics and Policy, Vol. 20. Peter Lang GmbH. Europaischer Verlag der Wissenschapften, Frankfurt am Main.

WUP (2001) *Review of Historic Water Resources Development and Water Use*. Working Paper No. 2 prepared by Halcrow Group Ltd in association with WRCS, Water Studies Pty, Finnish Environmental Institute, EIA Centre of Finland Ltd, and Team Consulting Engineers Co. Ltd, CamConsult Ltd, Laos Consulting Services, Water Resources University of Hanoi. Mekong River Commission, Phnom Penh.

9 Can We Maintain Food Production Without Losing Hydropower? The Volta Basin (West Africa)

WINSTON ANDAH,[1] NICK VAN DE GIESEN,[2] ANNETTE HUBER-LEE[3] AND CHARLES A. BINEY[1]

[1]Water Research Institute, Achimota, Ghana; [2]Zentrum für Entwicklungsforschung (ZEF), Bonn, Germany; [3]Stockholm Environment Institute, Boston, Massachusetts, USA

Introduction

The Volta Basin is located in West Africa. The main channel measures 1400 km and drains 400,000 km^2 of semi-arid and sub-humid savanna (Fig. 9.1). The basin lies mainly in Ghana (42%) and Burkina Faso (43%), with minor parts in Togo, Côte d'Ivoire, Mali and Benin. Ghana occupies the downstream part of the basin. A dominating feature of the basin is Lake Volta, which is the largest man-made lake in the world in terms of surface area (4% of the total area of Ghana). The lake was created to generate hydropower at Akosombo and Kpong (1060 MW) (Dickson and Benneh, 1987).

It is estimated that the country had a total population of about 18.9 million people in 1998, with an annual growth rate of about 3% per annum. About 65% of the total population in 1995 was rural, which had been 57% in a 1992 survey. A survey conducted by the Community Water and Sanitation Division (CWSD) of the Ghana Water and Sewerage Corporation (GWSC) in 1992 found that 57% of the population had access to a safe drinking-water supply. In rural areas 46% of the population, and in urban areas 76% of population, had such access.

The main land use in the Volta Basin is short bush fallow cultivation along the banks of the river network, and less intensive bush fallow cultivation elsewhere. Animal grazing is common, while the lake shore is extensively settled by fishing families. Charcoal production, involving the cutting of wood, becomes an extensive economic activity in the southern dry forest and transitional environments, e.g. the various parts of the Afram sub-basin on the west shores of Lake Volta. Since the 1960s, the Afram plains and other areas in the south, have been the focus of increasing settlement and agricultural development, having been generally thinly populated in the past (Dickson and Benneh, 1987). The forested and transitional areas are intensively farmed with

Fig. 9.1. The Volta River Basin; shared by Ghana, Burkina Faso, Ivory Coast, Togo, Benin and Mali.

cocoa, coffee, plantain, cocoyam, cassava, oil palm and maize on small bush fallow plots. Some timber extraction takes place in these areas. Recent developments, particularly below the Akosombo Dam, include irrigated rice, sugar and vegetable cultivation in the areas immediately adjoining the Volta River.

Water resources play a vital role in the promotion of economic growth and the reduction of poverty in Ghana. There is rapidly increasing demand for water in industry, particularly hydropower generation, agriculture, mining, recreation, domestic and industrial consumption, and environmental enhancement. With these demands, water supplies will be severely stretched and pollution problems and environmental degradation are likely to increase. The situation will worsen as the population continues to grow, urbanization increases, the standard of living rises, mining becomes increasingly widespread and human activities are more diversified. Lower rainfall amounts over the years due to longer dry seasons have led to more and more tributaries, as well as main rivers, drying up quickly, leading to lesser amounts of surface and ground water available for the increasing population.

Ghana is drained by three main river systems: the Volta, South-Western and Coastal River Systems. They cover 70%, 22% and 8%, respectively, of the total area of Ghana. The total annual runoff from all rivers is 56.4 billion m^3 of which 41.6 billion m^3 is accounted for by the Volta River (Table 9.1). The mean annual runoff from Ghana alone is 38.7 billion m^3, which is 68.6% of the total annual runoff. The Volta, South-Western and Coastal systems contribute 64.7%, 29.2% and 6.1%,

Table 9.1. Water resources availability in Volta Basin.

River basin	Area (km²)			% within Ghana	Mean annual runoff (×106 m³)			% within Ghana
	Within Ghana	Outside Ghana	Total		Within Ghana	Outside Ghana	Total	
Volta basin system								
Black	35,107	113,908	149,015	23.6	4,401	3,272	7,673	57.4
White	45,804	58,948	104,752	43.7	6,073	3,492	9,565	63.5
Oti	16,213	56,565	72,778	22.3	2,498	8,717	11,215	22.3
Lower	59,414	3,237	62,651	94.8	9,114		9,842	92.6
Total	**165,712**	**232,658**	**398,370**	**41.6**	**24,175**	**16,209**	**40,384**	**59.9**

respectively, of the annual runoff from Ghana. Runoff is marked by wide variability between wet season and dry season flows. Table 9.1 gives a summary of surface water availability within and beyond the country.

At the end of 1988, 12% of the total land area of Ghana was taken by cropland, 15% by permanent meadows and pastures, 36% by forest and woodland, while other land accounted for 37%. As of 1998, the total area cultivated under the four major starchy crops of the country (cassava, yam, cocoyam and plantain) increased from 1.21 to 1.31 million ha, while the total area cultivated for the major cereal crops increased from 1.28 to 1.34 million ha. The land under irrigation formed a negligible part of the arable and permanent cropland. Presently, only 10,000 ha out of a potential 346,000 ha are under irrigation.

Climate Change Projections

Four different climate change scenarios have been used: Hadley A2 and B2 and ECHAM4 A2 and B2. Figure 9.2 shows the temperature and precipitation projections for both Hadley and ECHAM A2. The values shown to the left of the dotted line are historical data for CRU and GCM. These GCM data have not yet been normalized. To the right of the dotted line, the normalized (read: 'downscaled', see Chapter 2) GCM projections are given for the simulation periods 1961–1990, 2010–2039 and 2070–2099. The data are summarized by so-called 'whisker boxes'. The horizontal line through the boxes gives the average value. The lower and upper sides of the boxes represent the average minus and plus one standard deviation, respectively. The 'whiskers' give the minimum and maximum values found in the set, unless real outliers (points that do not fit the general distribution) were present, which are marked by diamonds (MacDonald, 1992).

The temperature projections of the Hadley A2 GCM for the historical period compare well with the historical (measured) CRU data. Also for the other projections, this good comparison holds. An upward trend of 4.5°C/100 years can be seen with an increase in variability for the long-term prediction period. The GCM overestimates precipitation by 40% over the historical period. The normalized precipitation data

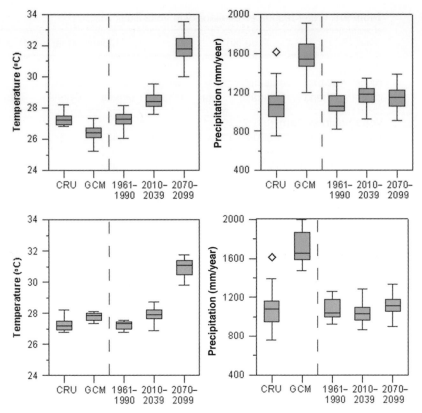

Fig. 9.2. Hadley (upper graphs) and ECHAM4 (lower graphs) A2 projections for the Volta Basin, showing future simulations for temperature and precipitation against measured data.

show a small but very relevant increase of almost 10%, with a significant increase in variability over the long-term projection period. The temperature projections of the ECHAM4 A2 GCM for the historical period compare well with the historical CRU data. An upward trend of 3.6°C/100 years can be seen with an important increase in variability for the long-term projection period. The shape of this upward curve is less linear, or more exponential, than for the Hadley model, suggesting ECHAM4 predicts a larger long-term and smaller mid-term change than the Hadley model. The GCM overestimates precipitation by more than 50% over the historical period. The normalized precipitation data show no increase for the mid-term and a small increase of about 8% for the long-term. Interestingly, the model predicts a decrease in variability over the mid-term projection period.

The B2 scenarios show similar trends, but less pronounced. The temperature projections of the Hadley B2 GCM show an upward trend of 3.1°C/100 years can be seen with no major increase in variability. As one may expect, this temperature increase was less than under the A2 scenario. ECHAM4 B2 GCM shows an upward trend of 2.5°C. Both models overestimate precipitation by 40% over the historical period. The average precipitation trends for Hadley A2 and Hadley B2 are comparable.

Table 9.2. Average yearly inflow into Lake Volta (km^3) together with standard deviation and coefficient of variation of the respective simulation periods for Hadley A2 and Hadley B2.

Scenario	Period	Average	SD	CoeffVar
Historical		32.8	17.1	0.52
HA2	2020–2039	41.6	14.0	0.34
HB2	2020–2039	43.8	15.4	0.35
HA2	2070–2099	37.2	19.9	0.54
HB2	2070–2099	44.0	17.6	0.40

Impacts from Climate Change

The impacts of climate change on the natural resources of the region depend on the changes or modifications in the atmospheric circulation over the area induced by the changes in global atmospheric chemistry. Climate change can have very serious negative effects on the socio-economic development of the country if the potential impacts are not identified for appropriate adaptive measures to be put in place. A number of sectors can be affected by climate change and these include water resources, coastal zone resources, agriculture, human health, energy, industry, forestry, fisheries and wildlife. Agriculture in Ghana is mainly practised under rain-fed conditions. Irrigation is practised in the basin on a very small scale. However, with increase in population and the need to meet food security under Ghana's poverty alleviation strategy, more land is envisaged to be put under irrigation.

Hydrology

As indicators of the impact of climate change on the hydrology of the basin, we use average annual runoff into Lake Volta and the variability in annual runoff as expressed by the coefficient of variation (= standard deviation/mean). A dry year is defined as a year with less than 75% of the average runoff during the period under consideration.

Table 9.2 and Fig. 9.3 show the inflow into Lake Volta for the historical and future periods using the WEAP hydrological model (WEAP, 2002). WEAP used both the A2 and B2 climate change scenario as input. The results show an important increase in runoff under each scenario. The underlying relative increases in rainfall were much less, but the non-linear response to slight absolute increases in rainfall causes the dramatic rise in runoff into the lake. The Hadley A2 simulation for 2070–2099 shows less increase and even a slight decrease in the coefficient of variation. Interestingly, the other simulations show an important decrease in the coefficient of variation. The B2 scenarios show in general a more important increase than the A2 scenarios, something that does not concur with the general idea that increase in temperature (which is higher under A2 than under B2) accelerates the hydrological cycle.

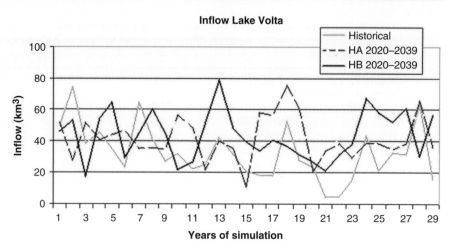

Fig. 9.3. Comparison of modelled flows for mid-term (2020–2039) and historical (1961–1990) periods using output from the Hadley A2 and B2 models.

Human health

Irrigation development creates a wide range of conditions that allow the growth and spread of disease vectors for malaria, schistosomiasis, etc. In general, these negative effects may be offset by improved living conditions and nutritional status. The construction of the Akosombo Dam eliminated the regular annual flooding of the Lower Volta River. This resulted in the development of sandbars that virtually blocked the estuary and prevented the upstream flow of sea water such that the estuarine salt wedge reduced from 30 to less than 5 km. Without the influence of saline waters, the clear waters of the Lower Volta became an ideal habitat for aquatic weeds and vectors of water-related diseases, particularly schistosomiasis. However, the impact of these activities on human health is difficult to estimate, so we have not defined a disease-related indicator (ODA, 1992).

Safe drinking water is also difficult to model as a function of climate change and adaptation strategies. An obvious indicator, such as percentage of households with safe drinking water, would be simply predicted by definition of the adaptation strategy ('increase of households with safe drinking water from 40% to 70%') and would, thereby, be meaningless.

Nature

Increase in irrigation and hydropower facilities for food production and cheap power generation could impact the environment negatively. These impacts include loss of the actual wetlands as they are modified to suit particular irrigation practices and associated losses in biodiversity. These impacts can eventually lead to socio-economic hardships as local people are deprived of useful wetlands. If, for example, the Bui Dam is constructed in the future, positive impacts will include increased availability of hydropower for industrial and urban development. The major negative impacts will be those associated with loss of the forest reserves, which are particularly rich in biodiversity, including a wide range of animals. The construction of the Akosombo Dam for hydropower, which drastically changed the annual flooding downstream, also resulted in the loss of several lagoons and creeks in the estuary that served as important fishing grounds. Associated with the shift in the estuarine salt wedge was the loss of the clam and prawn fisheries in the main channel that were a major source of livelihood (MWH, 2000, 2001).

As indicator for the biophysical impact of climate change, or associated adaptation strategies, we choose the number of *hectares of wetland lost*. The total wetland in the basin now is estimated at 2 million ha. The climate predictions show generally wetter conditions in the Volta Basin, which implies there will be no loss of wetlands under projected climate change. The wetland indicator is still relevant for measuring the impact of adaptation strategies.

Food

Ghana depends on agriculture for about 60% of gross national product and it provides work for about 80% of the population. Meanwhile, agriculture is almost wholly on a rain-fed basis. With an increasing population and growing demand for food, the economy can no longer depend on rain-fed agriculture. For the past 20 years, Ghana has experienced drought periods and erratic rainfall. Irrigation is therefore the way forward in agriculture if the country is to solve the food security problem. This is especially so in northern Ghana and Burkina Faso, where the rainy season lasts for only 4 to 5 months. The formal irrigation projects in Ghana (about 22 of them), which are sometimes referred to as large-scale irrigation, are almost overwhelmed with problems of finance, operation, management, etc. and therefore their impact is not notable in the economy. Small-scale irrigation development in the inland valleys, where one does not need huge initial capital, seems to have good potential in the short term. Irrigation development therefore will be very important in the economies of the countries in the basin.

As indicators for the food production sector, the *total tonnage produced* is used for two important crops, rice and maize. Rice is taken as representative for irrigated crops and maize for rain-fed crops. The SWAT field model was used to simulate crop productions for these two crops under the A2 and B2 scenarios. The Hadley A2 scenario simulation showed an increase in average yield from 3249 kg/ha to 3903 kg/ha and to 4688 kg/ha in the future under climate change (Fig. 9.4). The trend is similar for Hadley B2.

Maize on the other hand did not show much increase in the medium term (2010–2039) and rather a decrease in the long term for the Hadley A2 scenario. In the case of Hadley B2 there was virtually no change in yield.

As an input into the basin modelling, the yields in the SWAP field-level predictions in kg/ha were converted into production values by multiplying by the area under cultivation (FAO, 2003) for rain-fed and irrigation conditions (Table 9.3).

Industry (energy)

There is a continuously increasing demand for energy. The pressure to produce more energy is so high that the Volta River Authority (the energy-producing institution) lets

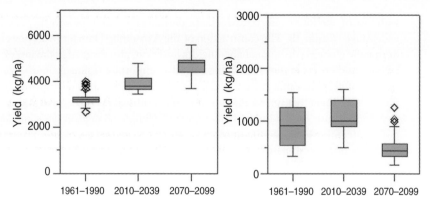

Fig. 9.4. Baseline yield projections for rice (left) and maize (right) under Hadley A2.

Table 9.3. Baseline crop production in tonnes.

	1961–1990	2010–2039	2070–2099
Rice	71,478	85,866	103,136
Maize	833,373	956,536	440,510

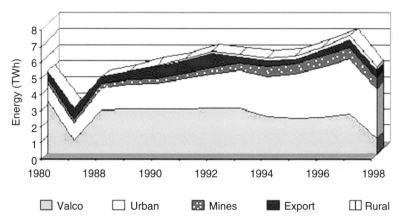

Fig. 9.5. Electric power use in Ghana. (Source: GLOWA, 2001.)

too much water through the dam with the hope that next year's rains will replenish the reservoir. When the rains are not so good for a single year, such as happened in 1997/98, there is no buffer and hydropower production comes to a halt. At present, hydropower generation is not very sustainable. The withdrawal rates are higher than the average inflow, leading to periodic shortages. The average annual generation is about 20,300 GJ and a graph of major users is presented in Fig. 9.5.

The Akosombo reservoir experienced two drought spells during the years 1983/84 and 1997/98, resulting in levels reducing below the minimum operating level (Fig. 9.6). As indicators for energy production we used the average energy production

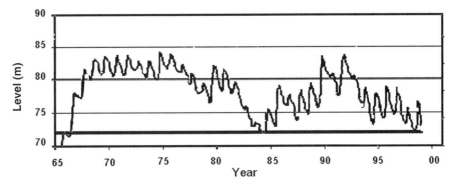

Fig. 9.6. Water level of Lake Volta. (Source: GLOWA, 2001.)

Table 9.4. Projected average energy production at Akosombo using the Hadley model for scenarios A2 and B2 and the historical range.

Scenario	Hydropower (GJ)
HA2 10–39	21,595
HA2 70–99	18,742
HB2 10–39	22,177
HB2 70–99	21,809
Historical 1961–1990	18,116

and the frequency of years in which Lake Volta has water levels below the minimum operational level. The latter result in lack of power, at least during parts of the year.

Future climate predictions show increased rainfall, which would benefit hydropower generation. However, in addition to the increased rainfall there is also increased variability, because the standard deviation also increases. WEAP simulations (Table 9.4) show that in general average energy production increases due to the increase in runoff. The only exception is the Hadley A2 long-term scenario (2070–2099), under which results are similar to historical conditions.

Figure 9.7 shows the impact of climate change for individual years. The figure shows that there is a definite decrease in years of failure. With the exception of Hadley A2 2070–2099, there are no years with power failures any more. Under Hadley A2 2070–2099, there are still 6 years with lake levels below the minimum level of operation for part of the year, comparable to the 7 found under the historical simulations.

Adaptation Strategies

Basin level

Three adaptation strategies were developed at basin level: business as usual, food security and energy. In business as usual, the present water use levels were maintained to assess what the isolated impact of climate change would be. Both future climate

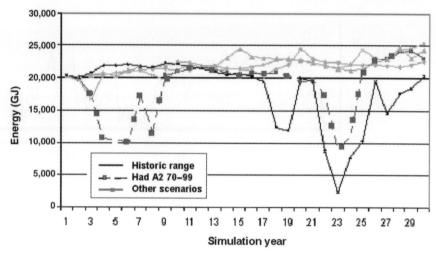

Fig. 9.7. Energy production at Akosombo for the different simulation periods using the Hadley model and A2 and B2 scenarios.

scenarios (Hadley A2 and B2) predict higher rainfall for the Volta Basin. It is not realistic to assume that existing infrastructure will be removed to favour environmental water flows. One could, therefore, see the business as usual adaptation strategy also as the best possible environmental adaptation strategy (CSIR-WRI, 2000).

The second adaptation strategy is the food security strategy. Food production is mainly rain-fed in the Volta Basin but irrigated agriculture is growing quickly. It is difficult to increase the productivity of rain-fed agriculture mainly because investments in labour or agro-chemicals do not pay off when the rains fail. Irrigation is therefore definitely an important means by which local food production can be increased. The present level of irrigation is very low. If Asian-type land use pressures existed in the Volta, probably more than 4% of the total area would be under irrigation. This would correspond with a 100-fold increase with respect to present levels, which is not realistic under present institutional arrangements. Instead, an irrigated area of 1% of the total area is put forward as the irrigation scenario, still an enormous relative increase from the present situation. We expect that most irrigation expansion will take place as small-scale, village-level irrigation.

The third adaptation strategy is the energy strategy. At present, the dam at Akosombo is used at unsustainable rates. The pressure to produce more energy is so high that the Volta River Authority (the energy-producing institution) lets too much water through the dam in the hope that next year's rains will replenish the reservoir. When the rains are not so good for a single year, such as happened in 1997/98, there is no buffer and hydropower production comes to a halt. In the business as usual scenario, we used a throughflow of 983 m^3/s, which is equal to the average flow over the past decades. In that case, flow demands were only unmet in 1983/84 under the historical scenario, which corresponds with actual disaster years in Ghana in which many forests burned down due to the drought (1983 also was the strongest El Niño year on record). In recent years, outflow levels of 1350 m^3/s were reached, which are not sustainable. For the energy scenario, it was assumed that the operation engineers will continue searching for more sustainable energy management measures. In practice, this translated into letting all extra runoff water that becomes available under projected future climates run through the Akosombo turbines.

The three adaptation strategies are relatively simple, but it can be stated with confidence that they span most of the realistic 'management space'. The business as usual adaptation strategy is definitely the minimum level of water resource development. The 1% irrigation adaptation strategy implies a very large extension of present irrigated areas and, as such, a maximum in irrigated area to be expected. Perhaps the only somewhat conservative strategy is the energy strategy because it assumes a reduction of the flows through Akosombo. It is, however, the only sustainable energy adaptation strategy because letting more water out from a reservoir than flows in clearly means that one runs out of water quickly.

Additional adaptation strategies that may be developed in the future could include a scenario in which the flows through Akosombo are concentrated in a shorter period in the wet season so that salt is allowed to enter the estuary as used to be the case before building the dam. It has been reported that the lack of movement in the salt front over the year has been the main environmental impact of Akosombo, destroying clam fisheries and increasing schistosomiasis. Thermal power could be used to bridge the dry season energy gap, but would most likely come at considerable financial costs. A second scenario that should be tried in the future is building the Bui Dam. It has many times been proposed to build a dam in the Bui gorge on the Black Volta, which would flood an almost uninhabited flood forest. This would be associated with unknown environmental damage. At present, it seems that a fully fledged Bui Dam would produce electricity at a cost of $0.09/kWh, whereas a thermal plant would produce at $0.07/kWh. A second alternative would be a smaller Bui Dam that would produce at economical costs but may not be interesting from a development-political point of view. Unfortunately, because the Bui Dam is presently in a bidding stage, no information concerning the properties of the dam and reservoir are available.

Field level

For two crops, rice and maize, three adaptation strategies at field scale are explored that address a baseline strategy mainly relying on rain-fed agriculture. Two other adaptations focus on irrigated and intensified conditions. The adaptation strategy mentioned as intensification consists of a package of cultivation practices, which in general will increase crop yields, e.g. improved crop variety, denser planting, chemical inputs and shorter season. Since rice is grown in the wet period, it was explored what the result of growing rice under only rain-fed conditions would be. Irrigated rice was, therefore, the baseline. Rain-fed rice was one adaptation strategy and intensified rice production was the second strategy. Maize production during the wet season is normally done under rain-fed conditions and this practice is used here as the baseline. The adaptation strategies for maize were growing irrigated maize and growing maize with intensified cultivation practices.

Under the baseline – rain-fed – adaptation, rice yields increase slightly for the medium term (2010–2039) under Hadley A2. However, in the long term the yield experienced not only a decrease but also high variability. Under intensification in its

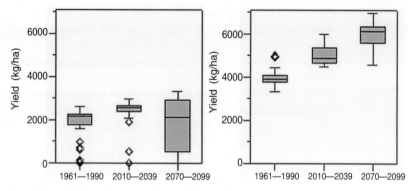

Fig. 9.8. Rice yield prediction under Hadley A2 for no irrigation and intensification.

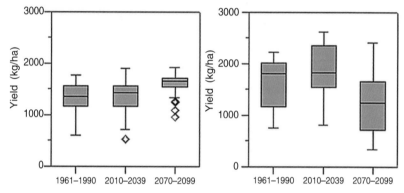

Fig. 9.9. Maize yield prediction under Hadley A2 for irrigation and crop intensification.

Table 9.5. Crop production under adaptation for maize and rice in tonnes.

	Adaptation strategy	1961–1990	2010–2039	2070–2099
Rice	No irrigation	40,216	53,196	39,830
	Intensification	8,734	110,242	131,186
Maize	Irrigation	1,186,680	1,192,973	131,186
	Intensification	1,476,158	1,675,736	1,087,790

cultivation there is a dramatic increase in the yield both in the medium and long terms. Intensification here means: using improved seeds, good agronomic practices, good soil fertility and pest and diseases management but without irrigation (Fig. 9.8). Maize yield under Hadley A2 showed a modest increase with decrease in variability in the long term under irrigation. Under intensified production, however, there was a high variability in the yield, which slightly increases in the medium term and unexpectedly decreases in the long term (Fig. 9.9) (Table 9.5) (MWH, 2001).

Table 9.6. Impact matrix for the main adaptation strategies and the Hadley A2 scenarios.

Volta	Indicator	Unit	Current	Future no adaptation	Future no adaptation	Adaptation			
			1990	2030 A2	2070 A2	2030 Food	2070 Food	2030 Energy	2070 Energy
Environment	Wetland loss	Ha	0	0	0	382,500	382,500	0	0
Food	Rice	Tonnes	71,478	85,866	103,136	2,004,400	2,385,200	85,866	103,136
	Maize	Tonnes	833,373	956,536	440,510	1,476,158	1,675,736	956,536	440,510
Energy	Energy/year	TJoule	18,116	21,595	18,742	21,121	18,095	23,232	19,467
	Low level	Years/30 years	7	0	6	0	7	8	10

Assessment of Basin-wide Adaptation Strategies

Table 9.6 shows the impact matrix for the different adaptation scenarios. If the wetland loss is acceptable, the food adaptation strategy seems relatively optimal. Remarkable also is the fact that even under increased runoff, it is not wise to rely totally on hydropower because there will still be regular periods where the water level is too low (Anonymous, 1992).

The interviews with experts at the national level, as well as recent literature, underlined the high priority for the electric power sector in Ghana. It is likely that this sector will develop various projects (hydropower, thermal plants, West African Gas Pipeline) in order to reduce the regular power shortages in the country. Supplying the population with sufficient electricity is at the top of the political agenda.

Decision makers in various Ghanaian institutions agreed that competing water use between hydropower and irrigated agriculture should be avoided. If irrigated agriculture is to be expanded, this should be in places where there is no competition between agricultural and hydropower use, as in the area below the Akosombo Dam. On the other hand, riparian countries also aim at expanding irrigated agriculture and developing hydropower. Burkina Faso, for example, is planning a hydropower dam for the supply of Ouagadougou. Togo and Ghana had an exchange agreement on electricity. Its basis is importing energy from Ghana in the rainy season and exporting it to Togo in the dry season. Togo furthermore exported energy to Ghana during peak times in daily power consumption. Togo has recently developed its own thermal plants and no longer depends on Ghana for power production. Close cooperation is needed for the riparian countries in the Volta Basin to prepare for further adaptation to climate change and climate variability.

References

Anonymous (1992) *Strategic Plan.* Environmental Protection Agency, Republic of Ghana, Accra.

CSIR-WRI (2000) *Climate Change Vulnerability and Adaptation Assessment on Water Resources of Ghana.* Water Research Institute. EPA, Accra.

Dickson, K.B. and Benneh, G. (1987) *A New Geography of Ghana,* 3rd edn. Longmans, Accra.

FAO (2003) *World Agriculture: Towards 2015/2030. An FAO Perspective* Bruinsma, J. (ed.) FAO/ Earthscan, London.

GLOWA (2001) *Global Change in the Hydrological Cycle (GLOWA).* GLOWA Volta Annual Report. ZEF, Bonn.

MacDonald, M. (1992) *Sub-Saharan Hydrological Assessment of West Africa.* Ghana Country report. MWH, Accra.

MWH (1998) *Water Resources Management Study: Information in the Volta Basin System,* Vol. 2. Ministry of Works and Housing, Accra.

MWH (2000) *Ghana Water Resources: Management Challenges and Opportunities.* Ministry of Works and Housing, Accra.

MWH (2001) *Agricultural Extension Policy,* Final Draft. Ministry of Food and Agriculture, Accra.

ODA (1992) *Environmental Synopsis of Ghana.* International Institute of Environment and Development, Accra.

WEAP (2002) Water Evaluation and Planning System. Available at: http://www.seib.org/weap

10 Will There be Sufficient Water under Internal and External Changes? Walawe Basin (Sri Lanka)

H.M. Jayatillake[1] and Peter Droogers[2]

[1]Irrigation Department, Colombo, Sri Lanka; [2]FutureWater, Arnhem, The Netherlands

Introduction

General context

The island of Sri Lanka is located south of India (Fig. 10.1) and is experiencing serious water resource problems. Walawe Basin (250,000 ha), located in the southern part of Sri Lanka, is inhabited by about 600,000 people, most of them dependent on agriculture, including fisheries. The basin's ambitious development plans indicate that the dominancy of agriculture in the region will change towards more industrial and service-oriented activities. These changes will have impacts on society as well as on natural resources, including water managerial issues. Currently, almost all the water resources are diverted for irrigated agriculture, with only a few per cent for industry and drinking water. Most recent development plans show that the use of water for urban use and industry will rise ten- to 15-fold (UNESCO, 2003).

Sri Lanka has a longstanding tradition of water management and the historical background of the country has helped create the perception that water is a public good. Agriculture is a tradition, in addition to being a major component of the livelihood strategies of the population. A local management structure for water resources was developed which included provisions for cost recovery and regulation. These provisions enabled a self-sustaining rural agrarian society to exist in the villages, but this is under threat as water issues become increasingly influenced at a basin scale.

The major natural hazard in the basin is drought. Monsoonal rains during the Maha season (November to March) and the Yala season (May to September) contribute a major part of the annual rainfall, which is supplemented by inter-monsoonal rains. The amount of rainfall reduces from the upper reaches to lower reaches and from west to east. The recent rainfall records at selected stations show a trend of

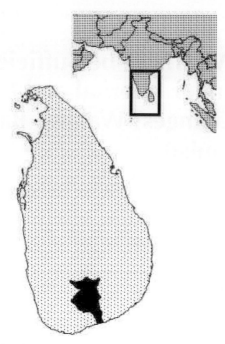

Fig. 10.1. Location of Walawe Basin in Sri Lanka.

decreasing annual rainfall since 1970. The ambient temperatures in the lowlands range from 25 to 28°C and in the upper elevations from 23 to 25°C.

Similarly, threats to the environment through pollution could increase. The coastal lagoon systems in the basin are close to areas of high population density and further expansion of industries can be expected in these areas. The major challenge for the future is managing competing demands whilst protecting ecosystems.

Malnutrition remains a problem in Walawe even though production of rice exceeds consumption. The number of stunted, wasted and underweight children in the area remains comparatively high, although substantial improvements have been made during the last decade. High levels of poverty will affect the sustainability of water resources management interventions, leading to a vicious cycle where water and land degradation restricts income, thus leading to further poverty.

On top of these internal changes that will have an impact on Walawe's water resources are the expected changes in climate that might put even more stress on water resources. A severe drought from 2000 to 2002, resulting in island-wide power cuts due to reduced hydropower generation, is seen by many as a first indication of what is going to happen.

Physical characteristics

Walawe is located in three, locally defined, climatic zones: dry, intermediate and wet. These groups are identified using long-term average annual precipitation and classed as: less than 1500 mm, dry; between 1500 mm and 2500 mm, intermediate; more than 2500 mm, wet. Defining an area as dry, while annual rainfall is still 1500 mm, must be considered in the context that reference evapotranspiration is about 1700 mm. Especially during the inter-monsoon periods, when rainfall is limited, severe water shortages can occur, resulting in reduced, or no yield, insufficient supply of drinking water and energy crisis. Temperatures in Walawe are constant, with temperatures in the lowlands ranging from 25 to 28°C and in the upper elevations from 23 to 25°C throughout the year.

The topography of Walawe Basin can be considered as a classical one: mountainous, relatively wet, catchment areas with limited development and downstream flat areas with developed water resources. These lowlands consist of rolling or undulating plain dotted with few isolated hills, while the rivers originate from the southern slopes of the central highland massif at elevations of up to 2000 m. Three geomorphic regions can be recognized across the basin. The highland region, which is made up of a complex of hill and valley landforms, and a major scarp at an elevation of more than 1000 m. The upland region is made up of a highly dissected plateau and has an elevation of between 300 and 1000 m. The lowland region, which accounts for about 70% of the basin, has developed water resources, including complex systems of linked reservoirs.

The dominant land type in Walawe is Chena, the traditional shifting cultivation system in Sri Lanka where, after one or two crops, land is abandoned for several years. Rice is grown in the lowlands covering an area of about 30,000 ha. The land use type 'garden' is often referred to as Other Field Crops (OFCs) and includes all crops except rice. OFC is sometimes irrigated, but mostly outside the official command areas. Tea can be found only in the highland region.

Water management in Walawe is based on a complex system of reservoirs, locally called tanks, which are linked in parallel as well as serially. A substantial number of minor tanks have been abandoned, some smaller tanks are used and managed by farmer groups and some major tanks are operated by the irrigation department. The total amount of storage is estimated at 600×10^6 m^3, covering about 5000 ha. Samanalwewa and Uda Walawe are the biggest in the basin. Both of them are used for irrigation as well as power generation, with an installed capacity of 120 MW (Samanalwewa) and 6 MW (Uda Walawe). This hydropower generation is of paramount importance for the country, as about 70% of electricity consumed is supplied by hydropower.

An overview of some characteristics describing the state of surface water resources can be found in Table 10.1. Water seems to be abundant in Walawe, as can be seen from the total annual average rainfall of 1860 mm. At the same time, only 33% of this water reaches the surface water system and the remainder is evaporated directly by open water bodies, soil evaporation and plant transpiration. Water availability expressed per capita is often used as a measure of water stress in a country or region. A true minimum human need for water can be defined as the amount of water to maintain human survival and is approximately 3–5 l of clean water for drinking. Increasing this amount to about 20 l of clean water per capita per day can realize improvements in human health substantially (Esrey and Habicht, 1986). A recommended basic water requirement for human domestic needs is set at 50 l of clean water per capita per day (Gleick, 2000). Besides these water requirements for

Table 10.1. Key characteristics describing the state of surface water resources. Data represent the entire Walawe Basin and are based on long-term averages.

Area (km^2)	2,500
Population	637,000
Precipitation (mm/year)	1,840
Precipitation (10^6 m^3/year)	4,600
Surface runoff (10^6 m^3/year)	1,500
Surface runoff fraction (%)	33
Outflow to sea (10^6 m^3/year)	525
Outflow fraction from precipitation (%)	11
Outflow fraction from surface runoff (%)	34
Rainfall per capita (m^3/year)	7,200
Surface runoff per capita (m^3/year)	2,400

basic needs, water is required for other services. Most important are water for food and water for nature. An average of 1000 m^3 has been proposed as reasonable to meet these requirements (Gleick, 2000). At the same time, the UN has proposed that areas having less than 1700 m^3 per capita water supply per year should be considered as experiencing water stress (Revenga *et al.*, 2000). Data from Table 10.1 indicate that water stress in Walawe is moderate, since the total water availability, based on precipitation in the basin, is 2666 m^3 and based on surface water resources is 888 m^3 per capita.

Shallow groundwater is of paramount importance to many people in the basin for their domestic needs. It is interesting that seepage losses from canals, reservoirs and paddy fields are indispensable for maintaining water levels in shallow wells, and several cases have been studied where canal lining resulted in less water in these shallow domestic wells. Only about one-third of the population has access to a piped water supply, indicating the dependency of so many people on this fragile linkage between surface water management and shallow groundwater.

Walawe has two major nature reserves: Uda Walawe National Park, just north of the Uda Walawe reservoir, and the downstream-located wetlands including Kalapuwa lagoon. Besides these two designated areas of natural importance, many other areas of the basin serve as biotypes for wild species: plants, birds, reptiles and mammals. The paddy fields, for example, are wetlands with a high number of bird species and should be protected. In terms of water resources and sensitivity to changing environments, including climate change, special attention should be given to the Uda Walawe National Park and the Kalapuwa lagoon. Some detailed studies for adjacent lagoons have been published recently, indicating the sensitivity of these lagoons to changes in upstream water management (Stanzel *et al.*, 2002).

Socio-economic and institutional characteristics

Poverty remains a problem in Sri Lanka, especially in the rural areas. While average GDP per capita for Sri Lanka was US$850 in 2000, it is estimated that this was about US$600 for Walawe. However, the main part of this GDP originates from subsistence

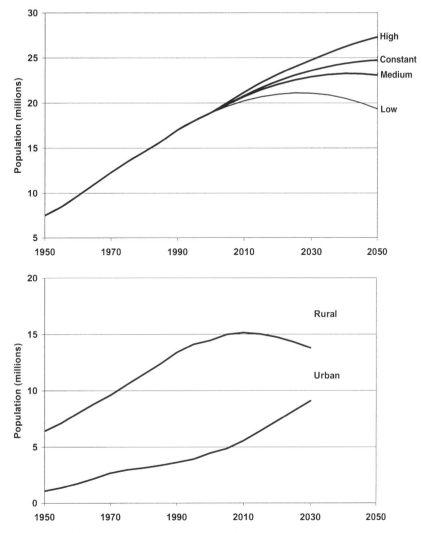

Fig. 10.2. Population growth and distribution between rural and urban areas. (Source: United Nations Population Division, http://www.un.org/esa/population/unpop.htm)

farming and average monthly household income is only US$50. Sri Lanka has a government-organized food and economic support system, from which nation-wide 39% of households benefit, while this percentage is 60% for Walawe.

A recent study from the World Bank (Anonymous, 2000) shows that poverty, expressed as household income, has been reduced in the period 1985–1990, but an increase in poverty was observed in the last decade. Only poverty of the urban population is reducing slowly. However, only 24% of Sri Lankans live in urban areas (UNFPA, 2002) (Fig. 10.2). Some key conclusions from the World Bank report are:

> . . . while Sri Lanka's per capita income remains higher than that of most of its South
> Asian neighbors, it lags far behind countries such as Singapore and Korea which had
> the same level of per capita income in the 1960s.   (p. 5)

> . . . missed opportunities for higher economic growth and poverty reduction are linked
> to Sri Lanka's failure to address priority issues including the ongoing civil conflict,
> inefficient government institutions, and constrained labor and land markets.   (p. 7)

A substantial number of agencies are responsible for water management in the basin. It is estimated that about 40 agencies deal with water at different levels, ranging from national to local. There are sectoral agencies dealing with domestic water supply, health and sanitation, agricultural and irrigation services, hydropower generation, groundwater development and ecosystem management. In addition, Provincial Councils are devolved with water-related functions. The Chief Secretary of the Province, District Secretary and Divisional Secretary are the key Government officials who make decisions regarding water resources management at their respective levels. At the district level, decisions on water are made by institutions such as the District Coordinating Committee, District Agricultural Committee and Project Management Committee.

Such a multitude of institutions calls for effective coordination at different levels. At the national level, the Central Coordinating Committee on Irrigation Management provides a forum for policy issues in irrigation management. A similar forum is the Steering Committee on Water Supply and Sanitation. The recent formation of Water Resources Council (WRC) addresses the need for coordination of water resources issues. Moreover, the formation of proposed River Basin Committees would address the existing inadequacy in addressing integrated water resources management issues.

The previously mentioned World Bank report (Anonymous, 2000) also urges that there is a need for major changes in the public sector:

> . . . growth in the number and size of Sri Lanka's public institutions is a major
> constraint to their efficiency. Today, politicization of recruitment by successive
> governments has been well established. Financial accountability has weakened due to
> insufficient Parliamentary oversight and poor institutional structures and watchdog
> bodies. The administrative apparatus remains very centralized.   (p. 9)

In terms of legislation it is estimated that over 50 Acts deal with the management of water resources. Implementing authority for these enactments is with a large number of agencies, dealing with different aspects such as irrigation, water supply, sanitation, industries and environment, as explained in the previous section. Public investment in water resources development concentrated on irrigation from 1950 until the 1980s, followed by a shift towards investments in rehabilitation and water management. However, in 2000 national investment in agriculture and irrigation was still about 8.5% of the total capital expenditure. The corresponding figures for energy and water supply sectors were about 16.5%.

The main investors in urban water supply and sanitation are the public sector, including Central Government, National Water Supply and Drainage Board (NWSDB), Provincial Councils and Local Authorities. Investments by community-based organizations and private individuals are significant in rural areas. Besides these national investments, a substantial number of irrigation and water-supply projects are still foreign-funded. Major irrigation rehabilitation projects and comprehensive groundwater assessment projects are funded by donors.

There were several attempts to recover the costs of operation and maintenance of irrigation services, which did not succeed. However, the ongoing programme of turnover of irrigation systems to the beneficiaries for management has enabled a substantial contribution from farmers to the system's management. The cost recovery mechanism for urban water supply focuses on operation and maintenance costs of the services. The level of recovery is lower in the water supply schemes managed by local authorities. The private investment at the family level in the construction of protected wells and latrines is considerable.

After a series of experiments from the late 1970s onwards, Sri Lanka adopted participatory management (in irrigated agriculture) as a state policy in 1988. A programme to turnover the management of irrigation systems to farmer organizations is ongoing. Although completely turned-over irrigation systems do not exist, there has been a significant improvement in the opportunities for farmers to participate in the management of irrigation systems in the last two decades. Most of the major irrigation systems in the basin are partially turned over, and minor irrigation systems (affecting an area less than 80 ha) traditionally have been managed by farmers.

The subject of climate change is virtually absent, at least explicitly, from policies prepared by the Government. This may be due to the fact that the phenomenon became widely appreciated only recently, and even then it failed to get attention since it was felt that it would mainly affect the industrialized nations.

Under the United Nations Framework Convention on Climate Change (UNFCCC), an Initial National Communication (INC) was made by the Ministry of Forestry and Environment, giving due recognition to the climate change phenomenon. The INC identified the issues arising from climate changes and proposed strategies to counter or mitigate the impacts. INC states that policy will have to focus on two aspects: reduction of emissions and mechanisms to mitigate impacts by following suitable adaptation strategies. Even though measures to cut down emissions are proposed, because the island is not a significant source of greenhouse gases (GHGs) such measures may not have an impact globally. In addition, INC recommends that the existing policies consider the impacts of climate change, when they are revised, and all new policy formulations allow for climate change impacts.

Climate Change Projections

The Intergovernmental Panel on Climate Change (IPCC) provides through its data distribution centre (http://ipcc-ddc.cru.uea.ac.uk/) results from seven General Circulation Models (GCMs). Somewhat arbitrarily, we have selected to use the model from the Hadley Centre for Climate Prediction and Research, referred to as HadCM3, and one from the Max Planck Institute für Meteorologie, referred to as ECHAM4. Since the GCM provides output at a low level spatial resolution, downscaling to local conditions was essential. Details about the downscaling of GCM results to the seven basins is described in detail in Chapter 2. For the historical data

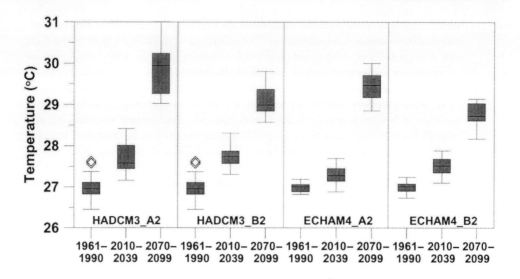

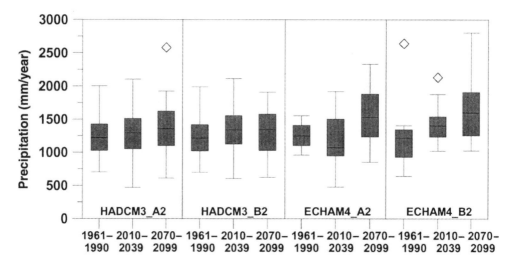

Fig. 10.3. Projected climate changes according to the two GCMs and the two Special Reports on Emissions Scenarios considered.

series, the East Anglia Climate Research Unit (CRU) database was used to provide data on temperature and precipitation for all seven basins over the 1961–1990 time slice (New *et al.*, 2000). This database provides a consistently interpolated global land surface dataset, with for each month between 1901 and 1996 an average value on a 0.5° × 0.5° grid.

From the various existing statistical transformations that ensure GCM and historical data have similar statistical properties, we used the method developed by Alcamo *et al.* (1997). For temperature, absolute changes between historical and future

GCM time slices are added to measured values, while for precipitation, relative changes between historical and future GCM output are applied to measured historical values. More details about this approach can be found in Chapter 2.

Figure 10.3 shows that for the near future (2010–2039) a small increase in temperature can be expected of about 0.5°C, while for the distant future (2070–2099) this increase will be between 2 and 3°C. The two GCMs considered are both showing the same trend, although HadCM3 projections are a little bit higher than the ECHAM4 ones. It is interesting that, especially for the A2 scenarios, an increase in variability can be expected.

Precipitation projections are less clear, but the overall trend is a somewhat wetter situation and simultaneously more variation in annual precipitation. However, HadCM3 shows an increase in this variation in the near future, but a decrease in the more distant future.

Impacts from Climate Change

Modelling framework

Climate change and adaptation strategies should be evaluated on different spatial scales. Since food production is a critical issue in Walawe, the field-scale model SWAP (van Dam *et al.*, 1997) has been used to deal with water and food issues. To consider these field-oriented analyses in a broader context and to include environmental issues, the basin-scale model WSBM (Droogers *et al.*, 2001a) has been set up for the basin.

The SWAP (Soil–Water–Atmosphere–Plant) is a one-dimensional physically based model for water, heat and solute transport in the saturated and unsaturated zones, and also includes modules for simulating irrigation practices and crop growth. The water transport module in SWAP is based on the well-known Richards' equation, which is a combination of Darcy's law and the continuity equation. A finite difference solution scheme is used to solve Richards' equation. A detailed description of the model and all its components is beyond the scope of this chapter, but can be found in van Dam *et al.* (1997).

In terms of irrigation requirements and food production, rice is the most important crop in Walawe Basin. It not only provides food for the people in the basin, but also is an important crop for export to the capital Colombo. Normally two crops can be grown during the year, one at the Maha and one at the Yala season.

Since no detailed crop characteristics were available, crop yields were computed using a simple crop-growth algorithm based on Doorenbos and Kassam (1979). The basic assumption of the simplified crop production function is that actual yield is a function of potential yields and water stress:

$$\frac{Y_{\text{act},i}}{Y_{\text{pot},i}} = \frac{T_{\text{act},i}}{T_{\text{pot},i}} \tag{1}$$

where $Y_{\text{pot},i}$ and $Y_{\text{act},i}$ are the potential and actual yield for a specific year i, and $T_{\text{pot},i}$ and $T_{\text{act},i}$ the potential and actual transpiration for year i. Sometimes evapotranspiration is considered instead of just transpiration, since determination of crop transpiration alone is difficult. Doorenbos and Kassam (1979) expanded this approach by including the effect that the sensitivity of the crop to water stress during subsequent growing periods is not constant:

$$1 - \frac{Y_{act}}{Y_{pot}} = K_y \left(1 - \frac{T_{act}}{T_{pot}} \right) \tag{2}$$

where K_y is yield reduction factor (–) indicating whether a crop is sensitive (>1) or less sensitive (<1) to water stress. K_y can have different values for different growing periods y.

A main drawback of this approach is the tendency to estimate high yields during low T_{pot} (hence low solar radiation) periods, and vice versa. We have therefore followed the approach to adjust the Y_{pot} accordingly to this T_{pot}, as applied successfully previous (Droogers *et al.*, 2001b). The potential yield for a certain year is assumed to be a linear function of the real maximum potential yield as obtained during very favourable climate conditions and optimal farm management:

$$\frac{Y_{act,i}}{Y_{pot,max}} = \frac{T_{act,i}}{T_{pot,i}} \times \frac{T_{pot,i}}{T_{pot,max}} \tag{3}$$

where $Y_{pot,max}$ and $T_{pot,max}$ are the maximum crop yield and maximum transpiration during the period of 30 years as considered in this study.

Obviously, the option to use a detailed crop modelling approach would be preferred, but since detailed crop parameters were lacking we have used the simplified approach as indicated. It should be emphasized here that this simplified approach is used very often and has proven to provide reasonable estimates (Droogers and van Dam, 2004).

Crop production is affected by the air's CO_2 level. Photosynthetically active radiation (PAR) is used by the plant as energy in the photosynthesis process to convert CO_2 into biomass. It is important in this process to make a distinction between C3 and C4 plants (see also Chapter 3). A large number of experiments have been carried out over recent decades studying the impact of increased CO_2 levels on crop growth. The Center for the Study of Carbon Dioxide and Global Change in Tempe, Arizona (http://www.co2science.org) has collected and combined results from such experiments for different crops, including rice (*Oryza sativa* L.). According to 26 references for rice, average biomass increase was 31% and 47% for increases in CO_2 air concentrations of 300 and 600 ppm, respectively.

At the basin scale, the Water and Salinity Basin Model (WSBM) was set up to link field-scale issues to the entire basin and to connect food issues with environmental ones. The main objective of WSBM was to create a simple and transparent water accounting model, to be used for quick analysis of river basin processes (Droogers *et al.*, 2001a). The model focuses on extractions for irrigation and the associated return flows from these systems. The model also includes a simplified urban and industrial water extraction component. WSBM works in an object-oriented style and was set up in Microsoft Excel to support transparency and flexibility. The concepts of WSBM are similar to that of WEAP (WEAP, 2002), but to ease coupling with the field-scale model SWAP, we have elected to use WSBM instead of WEAP.

WSBM assumes that the river is divided into nodes with a reach defined between two successive nodes. Nodes are located at typical points in the river where stream gauges are present or output is required. Water extractions, or supplies, occur only in the reaches. Using this approach water flows along the river can be simulated by subtracting extractions, or adding supplies, from one node to get the value for the next node. As mentioned before, extractions are defined for urban, industrial and irrigation supplies. For both types of extraction the amount of water and the return flow as a percentage of the extraction must be specified. Obviously, values can be either real data or hypothetical values to explore the effects of different interactions. The whole model was set up to run with a monthly time-step and it was assumed that the response time of the basin was within 1 month, so no time lag in water between months occurs. The model itself was applied, tested, calibrated and validated successfully for a basin in Iran (Droogers *et al.*, 2001a).

The two models were linked where changes in basin hydrology and field-scale hydrology interact. The entire set up is a flexible system where changes and adaptation strategies can be evaluated directly.

Indicators

The generic methodology applied here allows for quantifying food and environmentally related impacts under climate change. Based on these impacts, stakeholders are able to develop and evaluate different adaptation strategies to alleviate negative impacts of climate change (OECD, 1994; Aerts *et al.*, 2003). In the iterative approach, climate change scenarios are used as input to simulation models in order to quantify the impacts of climate change on the water resources system of a river basin and, consequently, the closely related implications for industry, the environment, food production and security.

For this, it is important to define a representative set of indicators, which reflect the value over time of the water resources system for preserving food security and environmental quality. Hence, impacts are here defined as the change in the values of indicators (Table 10.2).

A set of four indicators for the Walawe Basin has been defined. Preserving food security is best represented by two indicators: yield (t/ha) and food production (t/year). Furthermore, since environmental quality is mainly related to a lack of water availability for nature (such as wetlands) and to fresh water supply for coastal ecosystems, it has been decided to use two environmental indicators: outflow to the sea in 10^6 m^3/year and years with low flow in %/30 years.

Food and environment

The impact of climate change can be considered as the baseline (or business as usual) scenario. It provides a reference for what will happen under climate change if no adaptation measures are taken. The simulated changes for the periods 2010–2039 and 2070–2099 are compared with the current situation represented by the period 1960–1990. For this, the SWAP model has been used for calculating

Table 10.2. Effects table to assess the impact of a certain adaptation strategy. Indicator values express the change as a percentage relative to the baseline period 1961–1990.

			Indicator			
	Adaptation		Food		Environment	
	Area (%)	Irrigation (%)	Quantity	Security	Quantity	Security
			(% change)		(% change)	
No adaptation						
2010–2039	+0	+0	6	−10	28	−7
2070–2099	+0	+0	27	−8	78	13
Food adaptation						
2010–2039	+10	+10	19	−8	−15	−20
2070–2099	+10	+10	42	−8	28	−3
Environment adaptation						
2010–2039	−10	−10	−9	−14	85	17
2070–2099	−10	−10	10	−11	137	17

changes in food indicators at the field scale, whereas the WSBM model has been run to quantify changes in hydrology and consequently in basin-wide food production and the environment.

Figure 10.4 shows the impacts of climate change on food security. Crop yields will increase as a result of the enhanced CO_2 levels and the somewhat higher precipitation. However, at the same time a substantial increase in variation in yields is expected. But overall, the impact of changes in climate on food production levels is positive.

Outflow to the sea, used as the environmental indicator here, shows a somewhat similar picture, although the variation is much higher than seen for crop yield and total production (Fig. 10.4). However, if we consider the number of years that the minimum flow requirements are lower than the defined required one of $100 \times 10^6 \, m^3$, the situation is expected to change positively. In the period 1961–1990, minimum flow requirements were not met in 26% of the years. This changed to 33% and 13% for 2010–2039 and 2070–2099, respectively.

Overall, the impact of climate change appears to be positive from a water resources, food production and environmental point of view for Walawe Basin. However, if we take a somewhat closer look at the extremes in yield and productivity, there is some reason for concern. It is known that farmers are very vulnerable to extremes and especially the number of consecutive years of low yields. Farmers might be able to cope with 1 low-yield year, but 2 or even 3 consecutive years might be hard. There is the tendency that in the future the number of consecutive low-yield years increases. Figure 10.5 shows clearly that a clustering of extremes can be expected and coping strategies will be required to deal with this.

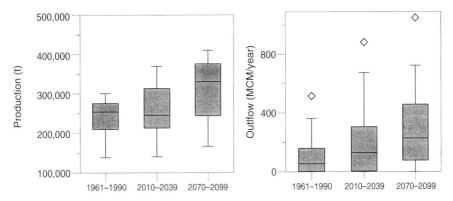

Fig. 10.4. Impact of climate change on total rice production (left) and outflow to the sea (right) using the HadCM3_A2 projection.

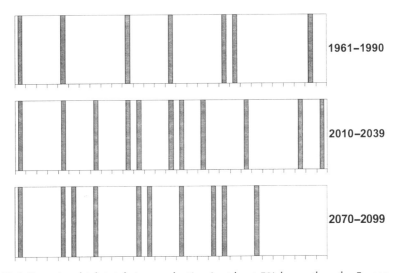

Fig. 10.5. Years in which total rice production is at least 5% lower than the 5-year moving average.

Adaptation Strategies

The overall impact of climate change on food production and environmental quality, as described in the previous section, appears to be positive. However, while the long-term average food situation is improving, variation in yield is increasing. Similarly, mean outflow to the sea is higher, but variation is also on the rise. Besides this externally driven climate change, the internal changes in the basin as described in the beginning of this chapter will put more stress on water resources, resulting in a reduced amount of water available for food production and the environment.

These external stressors include that more food will be required to feed the growing population in the basin as well as in the capital Colombo. There will also be increased demand for industrial domestic and environmental use. Simultaneously, there will be a growing pressure on land as more land is required for industry, urban use and service-oriented activities such as tourism.

To overcome these negative impacts of climate change and internal stressors on water resources, adaptation is required. Several attempts to generate a list of adaptation strategies have been undertaken, but no conclusive strategies have been developed. A nice example showing that defining adaptation strategies is not straightforward is shown by the Canadian Climate Impacts and Adaptation Research Network (CCIARN, 2002), which claims that 'priority setting' is the main issue. This was confirmed by a review of Smith *et al.* (2001), who stated that many different types of adaptation measures might be employed. Carter (1996) stated that possible adaptations are based on experience, observation and speculation about alternatives that might be created and they cover a wide range of types and can take numerous forms (UNEP, 1998). It is important to make a distinction between the different players as to who can do what (Droogers *et al.*, 2002).

The main problem facing the development of adaptation strategies is the level of detail included in the proposed strategies. Considering the problems expected in Walawe Basin it is clear that, although small and minor adaptation strategies might have some impact, we have to concentrate on basin-wide overall water resources issues. To address these issues two types of adaptation strategies can be considered: (i) a possible change in the total amount of water used for irrigation; and (ii) a change in the total cropped area in the basin. These two strategies were included in the overall analysis framework as developed for the basin and results will be discussed in the next section.

Evaluation of Impacts and Adaptation

The adaptation strategies will focus on two different priorities that policy makers can select: sustain food security and enhanced environmental quality. It is important to understand that we will not provide the ultimate solution, but we will provide output that shows what will be the consequences of a certain strategy. If it is decided that environmental quality has the highest priority, then our analysis will show how to realize this, how much quality will be gained, and what possible negative impacts there will be on other sectors. Similarly, if food security has the highest priority the framework will provide the available options and the consequences on, for example, the environment.

Food-focused adaptation

As discussed, the food-focused adaptation strategy is oriented towards two adaptation strategies: (i) changing the cropped area; and (ii) changing the amount of water delivered for irrigation. The change in cropped area is here implemented by only looking at the rice area, since rice is the dominant crop. The irrigation depth is adjusted

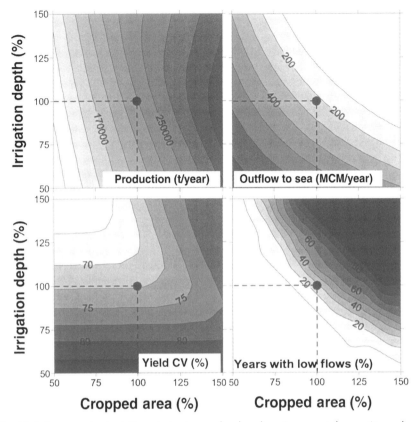

Fig. 10.6. Impact of adaptation strategies on food and environmental quantity and security using the baseline climate period 1961–1990.

according to water requirements. The model simulates whether the higher or lower demand can be fulfilled and, if not, adjusts. Actual levels are used for crop production calculations with the SWAP model.

The combined field- and basin-scale modelling framework has been run for different adaptation combinations of irrigation applications and cropped areas. Again, this has been done for the baseline period (1960–1990) to provide a reference, and for the near (2010–2039) and distant future (2070–2099). Figure 10.6 shows the effect of adaptation measures if future climate conditions are similar to those in the reference period 1961–1990. This graph provides information about food quantity (total rice production in the basin) as well as security (the combined variation in rice production between years, between Maha and Yala, and between different irrigation districts). The 100% value indicates the current situation, although since over-irrigation has been a common practice, the reality is more like 100% cropped area and 120% irrigation depth. The graph shows that for higher or lower irrigation depths, total production would remain similar. However, food security (expressed as the coefficient of variation between the 30 years, the different irrigation systems and the Maha and Yala seasons) would be lower with decreasing irrigation depths. The best option

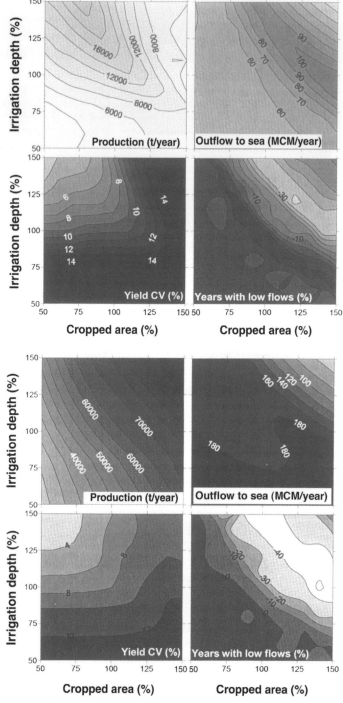

Fig. 10.7. Impact of adaptation strategies on food and environment quantity and security. Displayed are differences between the baseline period (1961–1990) and the near future (2010–2039, top) and the distant future (2070–2099, bottom).

to increase total production would have been to increase the cropped area, but this would have a negative impact on food security.

It is more interesting to look at the effect of the same adaptation options under climate change. Figure 10.7 shows the effect of adaptation strategies for the two selected periods of 30 years: 2010–2039 and 2070–2099. The most important conclusion is that compared to the business as usual scenario (100% irrigation and 100% cropped area), total production will be higher, especially in the distant future. However, the variation in production will increase from 72% in 1960–1999 to 82% in 2010–2039 and 80% in 2070–2099. It is up to water resources managers, but essentially to policy makers, whether such an increase is acceptable or whether intervention is required. Options, beside water managerial ones, are increasing the buffer capacity, by for example intervention prices, improved banking systems, social security systems, etc. The effectiveness of those options for alleviating climate change impacts, however, has not been explored in this project.

The graphs presented in the three figures can also be used to explore options in changing cropped areas and irrigation supplies. If we take the example of the increase in variation in production and have as a policy that an increase in variation, that is in food insecurity, is unacceptable, then the graphs can be helpful in exploring adaptation strategies. For the period 2010–2039, it may be concluded that cropped area should be reduced to about 75% of the current area and irrigation needs increase to a level of 130%. This will result in a variation in yield similar to the reference period, but total long-term production will be about 200,000 t/year, which is about 30,000 t less than the reference period.

The impact of increased water consumption by other sectors can be assessed in the same figures. If the expected increase in urban, industrial and service-oriented activities is 100×10^6 m^3, then diversions of water for agriculture should reduce by about 10%. The figures show that the impact on total production will be low, but variation in yield will increase.

The graphs presented are a useful tool for policy makers and water managers to explore the impact of climate change and climate variability and what kind of adaptation strategies can be developed towards food production and security.

Environment-focused adaptation

The previous section explored the impact of climate change and the effect of adaptation strategies on food production and food security. Similarly we have explored what can be done to enhance environmental quantity and security. Figures 10.6 and 10.7 show what the impact on the environment will be of the two most relevant adaptation measures related to water: (i) changing water diverted to agriculture; and (ii) changing the cropped area. As explained earlier, we have selected the total long-term outflow of the basin and the derivation from the defined minimum flow of 100×10^6 m^3/year to express environmental quality and security.

The reference period shows that currently about 200×10^6 m^3 flows out of the basin and the minimum flow requirement is not met in about 25% of the years. The graph can be used to explore what would have happened if the cropped area or the diversion to irrigation were different. However, it is more interesting to look at what will happen in the future if the irrigation depths and the cropped area remain unchanged. For the two time periods in the future, the overall impact on the environment will be positive. Total outflow will increase, and the number of years with low flow will increase somewhat for the 2010–2039 period, but decrease for 2070–2099. If this increase in low flow years is unacceptable, or if the policy is that low flows are only accepted in, for example, 10% of the years, then a decrease in the cropped area and/or a decrease in the water extracted for irrigation is unavoidable.

These environmental adaptation strategies should not be considered in isolation, but the effect on food should be included as well. In general, improving environmental quality and security by decreasing the cropped area will result in lower total production. Alternatively, decreasing the diversions to irrigation will have a minor impact on total food production, but will increase the variation in food production.

Again, it is the policy maker who should decide what the future should look like. The results presented here aim to help in making sound decisions.

Conclusions

It is clear that for Walawe Basin in Sri Lanka climate change will have an impact on water resources and therefore on food security and environmental quality. Precipitation will be higher, temperatures will increase, evapotranspiration demands will be higher and potential crop growth will rise due to elevated CO_2 levels. The overall impact of these factors is that long-term average yields and production will increase, but at the same time extremes will occur more frequently.

Socio-economic issues in Walawe Basin are expected to undergo major changes. The current agriculture-oriented society is likely to change to a more industrialized and service-oriented one, having impacts on water resources, food and the environment. On top of this will be the impact of climate change and changes in climate variability. In this research a comprehensive framework was developed to evaluate adaptation strategies for alleviating impacts by climate change. The framework uses current data, identifies relevant water resources issues, quantifies future trends, generates climate change projections and integrates model results to support water managers in strategic decision-making.

Overall, the threat on water resources issues in the basin as a result of these internal and external factors appears to be small. The expected increase in water required for industry, urban and service-oriented activities of about $100 \times 10^6 \, m^3$ is about 10% of the overall water consumption by irrigated agriculture. Crop production will increase as a result of the small increase in precipitation, but mainly as a result of the positive impact of enhanced CO_2 levels on plant growth. The major concern is the increased variation in climate. The basin has experienced some consecutive dry years over the last decade, and according to our analysis this will happen again in the future. Policy should therefore be directed to this and some possible options are mentioned in this document. A summary of these options is presented in the effect table (Table 10.2), and more details can be extracted from Figs 10.6 and 10.7. Obviously, more extensive details can be obtained by using the modelling framework as developed under this study.

The analysis framework strongly emphasizes the main issues related to water–food–environment in the context of Walawe Basin: rice production and irrigation. The physically based field-scale SWAP model provides detailed soil–water–plant processes, and the basin-scale WSBM assesses water allocation. The coupling of the two models ensures linkages between food production and the environment.

This study could be considered an implementation of the Initial National Communication (INC) as initiated by the Ministry of Forestry and Environment under the United Nations Framework Convention on Climate Change (UNFCCC). The INC provides the general structure and is the first step towards climate change policy in Sri Lanka.

Outstanding issues are adaptation options related to rainfall harvesting, rainfall–runoff processes, and, most importantly, hydropower generation. Indirectly, it has been shown that more water will become available in the future, but at the same time the variation will be greater, putting more pressure on the vulnerable energy situation in the country.

Another important factor is the dominant influence of CO_2 on potential crop production. Although the complex interactions between crop production and enhanced CO_2 levels are still not fully understood, the literature and experiments provide sufficient proof that yields could increase substantially.

Finally, the presentation of the output of this study is attractive for policy makers and water resources managers. More options can be assessed by looking at the detailed output, or by using the framework presented in this study for analysing other adaptation strategies.

References

Aerts, J.C.J.H., Lasage, R. and Droogers, P. (2003) *A Framework for Evaluating Adaptation Strategies.* Institute for Environmental Studies, report R-03/08. Vrije Universiteit Amsterdam, The Netherlands.

Alcamo, J., Döll, P., Kaspar, F. and Siebert, S. (1997) *Global Change and Global Scenarios of Water Use and Availability: An Application of WaterGAP 1.0.* Report A9701. Center for Environmental Systems Research, University of Kassel.

Anonymous (2000) *Sri Lanka Recapturing Missed Opportunities.* Report no. 20430-CE. World Bank, Washington, DC.

Carter, T.R. (1996) Assessing climate change adaptations: the IPCC guidelines. In: Smith, J., Bhatti, N., Menzhulin, G., Benioff, R., Budyko, M.I., Campos, M., Jallow, B. and Rijsberman, F. (eds) *Adapting to Climate Change: an International Perspective.* Springer, New York, pp. 27–43.

CCIARN (2002) *Research Priorities.* Canadian Climate Impacts and Adaptation Research Network. Available at: http://www.c-ciarn.uoguelph.ca/research_priorities.html

Doorenbos, J. and Kassam, A.H. (1979) Yield response to water. FAO irrigation and drainage paper 33. FAO, Rome.

Droogers, P. and van Dam, J.C. (2004) Field scale adaptation strategies to climate change to sustain food security: a modeling approach across seven contrasting basins. IWMI Working Paper. International Water Management Institute, Sri Lanka.

Droogers, P., Salemi, H.R. and Mamanpoush, A.R. (2001a) Exploring basin scale salinity problems using a simplified water accounting model: the example of Zayandeh Rud Basin, Iran. *Irrigation and Drainage* 50, 335–348.

Droogers, P., Seckler, D. and Makin, I. (2001b) Estimating the potential of rainfed agriculture. IWMI Working Paper 20. International Water Management Institute, Sri Lanka.

Droogers, P., Hoogeveen, J. and van Dam, J. (2002) Adaptation strategies to enhance food security. ADAPT paper. Institute for

Environmental Studies, Vrije Universiteit Amsterdam, The Netherlands. Available at: http://www.geo.vu.nl/users/ivmadapt/fb_ papers.htm

Esrey, S.A. and Habicht, J.P. (1986) Epidemiological evidence for health benefits from improved water and sanitation in developing countries. *Epidemiological Reviews* 8, 117–128.

Gleick, P.H. (2000) *The World's Water 2000–2001. The Biennial Report on Freshwater Resources.* Island Press, Washington, DC.

New, M., Hulme, M. and Jones, P. (2000) Representing twentieth century space-time climate variability. II: Development of 1901–1996 monthly grids of terrestrial surface climate. *Journal of Climate* 13, 2217–2238.

OECD (1994) *OECD Core Set of Indicators for Environmental Performance Reviews.* Organization for Economic Cooperation and Development, Paris.

Revenga, C., Brunner, J., Henninger, N., Kassem, K. and Payne, R. (2000) Pilot analysis of global ecosystems: freshwater systems. World Resources Institute. Available at: http://www.wri.org

Smith, J., Lavender, B., Smit, B. and Burton, I. (2001) Climate change and adaptation policy. *Canadian Journal of Policy Research* 2, 4–8.

Stanzel, P., Öze, A., Smakhtin, V., Boelee, E. and Droogers, P. (2002) Simulating impacts of irrigation on the hydrology of the Karagan Lagoon in Sri Lanka. Working Paper 44. International Water Management Institute, Sri Lanka.

UNEP (1998) *Handbook on Methods for Climate Impact Assessment and Adaptation Strategies.* Feenstra, J.F. *et al.* (eds) Institute for Environmental Studies, Vrije Universiteit Amsterdam, The Netherlands.

UNESCO (2003) *Water for People, Water for Life. United Nations World Water Development Report.* UNESCO, Paris.

UNFPA (2002) The state of the world population in 2001. United Nations Population Fund. Available at: http://www.unfa.org/swp/2001/english/indicators/indicators 2.html

van Dam, J.C., Huygen, J., Wesseling, J.G., Feddes, R.A., Kabat, P., van Walsum, P.E.V., Groenendijk, P. and van Diepen, C.A. (1997) Theory of SWAP version 2.0 Technical Document 45. Wageningen Agricultural University and DLO Winand Staring Centre, Wageningen, The Netherlands.

WEAP (2002) Water Evaluation and Planning System. Stockholm Environment Intitute, Boston, Massachusetts. Available at: http://www.seib.org/weap

11 How Can We Sustain Agriculture and Ecosystems? The Sacramento Basin (California, USA)

Annette Huber-Lee,[1] David Yates,[2] David Purkey,[3] Winston Yu,[1] Charles Young[3] and Benjamin Runkle[1]

[1]Stockholm Environment Institute Boston, Boston, Massachusetts, USA;
[2]National Center for Atmospheric Research, Boulder, Colorado, USA;
[3]Natural Heritage Institute, Berkeley, California, USA

Introduction

Water development in California is among the most extensive in the world, with water shifted from one basin to another over distances of hundreds of kilometres in order to satisfy water demands. The main river systems are the Sacramento and San Joaquin Rivers that converge in a region known as the Delta that then flows into San Francisco Bay – the combined area is referred to as the San Francisco Bay Watershed (SFBW). Below Fresno, the Central Valley is in fact a closed basin associated with what was once the Tulare Lake, although the lakebed itself has been reclaimed for irrigated agriculture through the impoundment and regulation of the rivers entering that portion of the valley (Fig. 11.1). Most water management issues relate to ever increasing urban demands competing with agricultural and ecosystem needs. Primary among the sources of water available in California is the Sacramento River, which not only supplies demands within the basin and the critical agricultural area of the Central Valley, but also supplies municipal and industrial demands on the Southern California Coastal Plain between Los Angeles and San Diego (Table 11.1). As such, the water resource situation in the Sacramento Basin cannot be discussed in isolation from the situation statewide.

It is anticipated that metropolitan regions in the Central Valley, such as Sacramento, will continue to grow dramatically in the future as the large coastal metropolitan areas, such as the San Francisco Bay area and Los Angeles, become increasingly crowded. This urban expansion is taking place beside large-scale agriculture, therefore competing with both land and water resources.

The Sacramento River and its tributaries convey 31% of California's average annual runoff, a water resource that has supported the development of over 850,000 ha of irrigated agriculture in the basin, as well as expansive irrigation development in

Fig. 11.1. Relief map of the State of California, with the location of the Sacramento River Basin.

other parts of the state. The principal crops grown in the Sacramento Basin include rice, olives, orchard fruits and nuts, maize, lucerne, tomatoes and vegetables. For many of these commodities, the basin is a globally important production centre.

In addition to urban and agricultural use, the waters of the Sacramento Basin also support several important ecosystems. Not surprisingly, the development of irrigated agriculture has dramatically changed the natural landscape in the basin. Today only 5% of historic wetlands in the Sacramento Basin remain, and these are contained largely within the 13,000 ha of the Sacramento National Wildlife Refuge Complex. In addition, only 5% of the original 200,000 ha of riparian forest along the river and its tributaries remains. While the Sacramento River and its tributaries continue to support some of the southern-most runs of Pacific salmonids, the continued viability of these runs is threatened by extensive water development. Three ecosystem components are of particular interest and are presented here, although many other important ecosystem services are provided. The first ecosystem component of note is the anadromous fishery, and most notably the Chinook salmon fishery that spends a portion of its life cycle in the Sacramento Basin. The second is the waterfowl migrating along the Pacific Flyway that rely upon wetlands in the Sacramento Basin during their north–south migration. The final ecosystem component of note is the riparian cottonwood and willow forests that shelter many birds and mammals in the Sacramento Basin.

The climate in the Sacramento Basin, as in much of California, is Mediterranean in character, typified by wet winters and dry summers. Most precipitation occurs during the period between November and April, with little or no precipitation falling

Table 11.1. Approximate Sacramento Basin water budget.

		Annual volume (million m³)
Sacramento Basin runoff	A	27,630
Import from Trinity River system	B	1,087
Total Sacramento Basin supply	C: A+B	28,717
Export to southern California	D	1,938
Export to Tulare Lake Basin	E	3,020
Export to San Joaquin Valley	F	1,806
Export to San Francisco Bay area	G	113
Unexported Sacramento Basin runoff	H: (A+B)−(D+E+F+G)	21,840
Sacramento Basin urban water use	I	871
Sacramento Basin agricultural water use	J	9,948
Unallocated Sacramento Basin runoff	K: H−(I+J)	11,021
As a percentage of Sacramento Basin runoff	L: K/A×100	40%

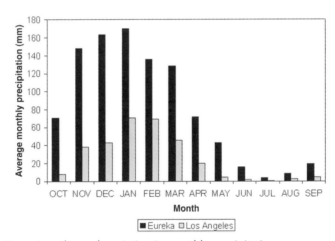

Fig. 11.2. Current north–south variation in monthly precipitation.

between May and October. In addition to the north–south variation, there is also an east–west variation in precipitation that is controlled largely by the orographic effect of mountains on Pacific storms (Fig. 11.2).

Given the climate conditions common to California, it is not surprising that the surface water hydrology of the Sacramento Basin is dominated by winter snowfall and subsequent spring runoff. Prior to the initiation of large-scale water development in the basin, this climate pattern resulted in flow maxima in the Sacramento River main stem and its principal tributaries – the Feather, Yuba and American Rivers – during the late winter through spring period. Water development in the basin, primarily the construction of major reservoirs on all of the major tributaries, has dramatically altered the surface water hydrology in the basin. The operation of these reservoirs

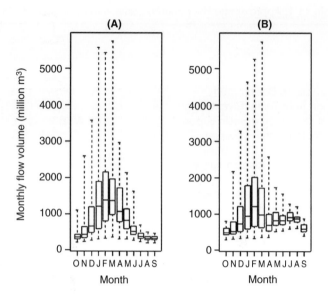

Fig. 11.3. Monthly flow volumes in the Sacramento River below Shasta dam as (A) an estimate of the full natural flow and (B) the observed flow.

generally creates peak flow conditions earlier in the winter as operators manipulate reservoir storage as part of flood control operations in advance of the main runoff season. Spring flows are typically reduced as operators attempt to capture reservoir inflow for later release as part of water supply operations. As a result, summer flows are significantly higher than under natural conditions as operators release water downstream to meet summer irrigation demands (Fig. 11.3B).

Currently, there are three distinct perspectives on the state of water management in the Sacramento Basin. The first is that the balance between water for food and water for the environment has been destructively tipped in favour of irrigated agriculture and that the only possible future is one based on constant efforts to roll back the irrigated area in the basin. The second is that the Sacramento Basin is too valuable an agricultural resource to be constrained by environmental considerations and that issues of water for the environment should be dealt with in other, less valuable, areas. Both of these views are increasingly giving way to a third perspective that seeks to balance the complex trade-offs and interactions between water for food and water for the environment in the basin. Establishing this balance is a work in progress, and the prospect of climate change offers the real possibility that the emerging balance will be upset and that further adaptation will be required.

Climate Change Projections

Climate change and increased climate variability may have a profound impact on the availability of water resources in the Sacramento Basin and could seriously affect the

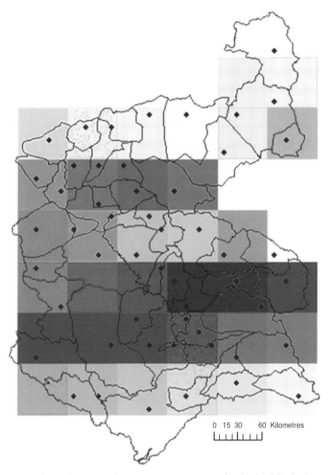

Fig. 11.4. GCM grid overlay over the Sacramento watershed. The black dots represent weather stations.

multiple uses of water of the Sacramento, including domestic, industrial, agricultural and ecosystems. The importance of understanding the trade-offs and interactions among competing water uses will only increase with the added impact of climate change. Relevant to the Sacramento Basin, GCM projections estimate that: (i) average temperatures could increase by as much as 5°C; and (ii) mean annual precipitation may decrease over the period 1990–2100. However, at their current resolution, GCMs drastically smooth out most of California's complex topography. For example, current GCMs do not contain important terrain features such as the Coastal Range, the Central Valley and the Sierra Nevada Range. Comparisons of climate change patterns for California that emerged from an analysis of 21 GCMs showed that all models estimated warmer temperatures for the state under assumptions of greater radiative forcing from increased greenhouse gas emissions (Gutowski *et al.*, 2000). For this project we evaluate impacts and adaptation strategies for the Hadley A2 and B2 scenarios. Figure 11.4 shows the 0.5° × 0.5° grid overlay for both climate scenarios over the Sacramento watershed.

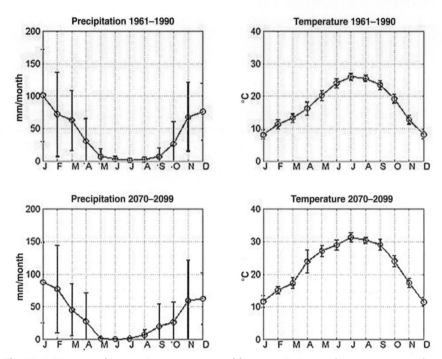

Fig. 11.5. Upper and Lower Sacramento monthly precipitation and temperature (with 2×standard deviation error bars) for the historical record (1961–1990) and future projections based on the HADCM2 GCM for 2070–2099.

The effects of climate change will vary across the Sacramento Basin. As an illustration, Fig. 11.5 shows the change in the monthly precipitation and temperatures for the periods 1961–1990 to 2070–2099 for two representative sub-basins, one in the Lower and one in the Upper Sacramento for the Hadley A2 GCM climate scenario. Two standard deviation error bars (standard deviation of monthly time series) are also shown and represent the monthly variation over the given time period (IPCC, 2001).

In both sub-basins, temperatures are expected to increase by about 5°C. On average, precipitation is expected to decrease, primarily during the winter months, contrary to the outcomes of the Canadian GCM but consistent with the Japanese GCM. The magnitude of this reduction is expected to be significantly larger for the Upper sub-basin than for the Lower Sacramento sub-basin. For the Lower Sacramento, the average total annual precipitation is 494 mm, 438 mm and 411 mm over the 1961–1990, 2010–2039 and 2070–2099 periods, respectively. For the Upper Sacramento, the average total annual precipitation is 1225 mm, 1062 mm and 1035 mm over these same periods, respectively.

Furthermore, examining the unadjusted precipitation time series, the coefficient of variation (CV) (ratio of standard deviation to mean) increases over the two time periods suggesting increased variability. The observed CV increases from 1.08 to 1.26 from the observed historical period to 2070–2099 for the Upper Sacramento. Similarly for the Lower Sacramento, the CV increases from 1.37 to 1.44. Lastly, analysis reveals that the persistence of anomalous climate events will also increase, the magnitude of which is larger for anomalous temperature events than precipitation events.

Impacts of Climate Change

In order to quantify the potential impacts of climate change, a representative set of *indicators* is defined to illustrate the changes in food production and ecosystem health over time, as proxies for the goals of food and environmental security.

Preserving food security is represented by changes in two indicators:

- agricultural production; and
- variation in annual agricultural production.

Note that for the purposes of this chapter, agricultural production is assumed to be a linear function of the crop water requirements. Therefore, unmet agricultural demands act as a measure of changes in agricultural production. Variation in annual agricultural production is defined as the standard deviation of the unmet agricultural demands over the time period of interest.

Environmental security is related to impacts on the natural ecosystems as well as impacts on human health and well being. It is represented by changes in the following four indicators:

- salmon population;
- wetland area, including area of rice flooded in the winter season;
- availability of water for domestic purposes; and
- aquifer storage.

Salmon population provides a measure of the overall instream aquatic health. We use as a proxy for changes in salmon population the change in frequency of unmet instream flow requirements designed to support salmon habitat. Wetland area is a measure of the riparian habitat; a unique ecosystem important for many wildlife species as well as providing water purification. Water for domestic consumption has direct bearing on human health and well-being. Finally aquifer storage gives an indication of the overall sustainability of the system.

Changes in the above indicators are examined against implementation of various adaptation strategies, including a 'business as usual' strategy, described below.

Business as usual (no climate change)

Even without climate change, the expected increase in population will lead to an increase in domestic demand as well as changes in land use (i.e. increased urbanization) (J.D. Landis and M. Reilly, 2003, unpublished paper). This will intensify pressure on water resources in the Sacramento watershed. Bulletin 160-98 of the California Water Plan Update estimates that at 1995 levels of development, water shortages already exist and are in the order of 2000 million cubic metres (10^6 m^3) in average water years for the entire state. In drought years, the shortage more than triples to 7000×10^6 m^3. By 2020, due to population-driven demand growth alone, it is estimated that the shortages will be in the order of 3000×10^6 m^3 in an average water year and 8000×10^6 m^3 in drought years for the state of California, and 105×10^6 m^3 and 1220×10^6 m^3 respectively for the Sacramento watershed (Department of Water Resources, 1998). The Sacramento is in part vulnerable to water shortages as substantial supplies are exported to meet demands in other parts of the state. An aspect of the future that has not explicitly been explored in the state so far is the impact of land use changes on the hydrology of the system, and in particular a shift of land use from agriculture to urban areas.

Model results show much higher shortfalls than the Department of Water Resources (DWR) Sacramento projections of deficit for an average year without climate change. Modelled shortfalls are 505×10^6 m^3. One possible explanation is that the DWR water budget did not consider exports from the Sacramento region to other parts of the state. The export from the Sacramento delta to the San Joaquin Valley and Los Angeles are currently modelled as major demands (i.e. with a combined urban and agricultural demand of approximately 7400×10^6 m^3). Another possible explanation for this difference is that the California state projections do not account for the impact of land use change on the basin hydrology as predicted by Landis and Reilly (2003, unpublished). This impact is included this study and is probably negatively impacted.

Of the 505×10^6 m^3 shortfall, agricultural demand accounts for 482×10^6 m^3, while the remainder is urban demand. Furthermore, in-stream flow requirements for the anadromous fish recovery programme (AFRP), particularly for the American River tributary, are consistently not met. On average, flow requirements in the month of July are not met 69% of the time. This is consistent with current conditions in this river.

Impacts of climate change: hydrology

Existing extreme variability in precipitation throughout the basin combined with high levels of demand make the Sacramento Basin particularly vulnerable to climate change. Based on runs with the WEAP (2002) model, the net impact on the annual flow in the upper reaches of the Sacramento under the Hadley A2 climate scenario is an 11% decrease during the period 2010–2039, with a further decrease to 24% in 2070–2099, relative to no climate change. This decrease in flow occurs primarily from February to July, as illustrated in Fig. 11.5 below, for both climate change periods. Note that for the Land Use Change scenario only, the climate used is that of the historical record (see also Fig. 11.6).

Climate change could make water supplies more vulnerable due to reduced snow packs and thus lower summer streamflows, which would be a threat broadly to water-related ecosystem services, including municipal and agricultural sectors, recreational and commercial fishing, recreational viewing, as well as overall ecosystem health.

Impacts of climate change: food security

The current system of priorities in the basin generally is first to supply ecosystem requirements, second urban needs, and finally agricultural needs. Both agricultural

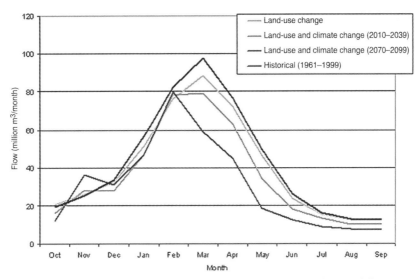

Fig. 11.6. Average monthly flows in the Upper Sacramento according to different scenarios and baseline situation (1961–1999).

and urban demands will increase under climate change, in spite of a shift of land from agriculture to urban use. This is due to the increased temperatures under climate change affecting evapotranspiration, which in turn leads to increased crops water requirements that outweigh the loss of agricultural area. Future average water demands are shown in Table 11.2. Environmental demands are assumed not to change in the future. Agricultural demand in all periods is in the order of 90% of the total water demand in the Sacramento system. Interestingly, the largest demand growth is in the urban sector (i.e. a factor of 4 over the next 100 years compared to a factor of 1.2 for agriculture). Still, agricultural demands clearly dominate the system. Agricultural demands are slightly larger under the B2 climate scenario due to less overall precipitation, offsetting the smaller projected temperature increases.

With this system of priorities, environmental demands are essentially satisfied under both the no-climate change and with-climate change scenarios. With land use changes only, both agriculture and urban areas face shortfalls in the order of about 5%. This unmet urban demand is entirely from Placer and El Dorado counties due to the higher priority downstream American River AFRP flow requirements. With climate change unmet demands increase over time for both agricultural and urban users.

Average unmet agricultural demand with the 'land use changes only' scenario (i.e. using the historical climate) is approximately 482×10^6 m^3 (standard deviation $= 329 \times 10^6$ m^3) for the period 2010–2039. With climate change, the average unmet agricultural demand increases to 786×10^6 m^3 (standard deviation $= 359 \times 10^6$ m^3) in 2010–2039 (see Fig. 11.7 below). On a percentage basis, unmet agricultural demand increases from 5% to 7% with climate change in 2010–2039, and to 12% in 2070–2099. This baseline 5% unmet demand reflects predominantly unmet demands for irrigated pastures and orchards in the upper watersheds of the Sacramento (e.g.

Table 11.2. Average sector demands for historical period and two future projected climate periods 2010–2039 and 2070–2099 (both are HA2 GCM scenarios). Ranges are given in brackets.

Sector	Historical period (1961–1999)	Projected period (2010–2039)	Projected period (2070–2099)
Agriculture (10^6 m³)	10,075 [7,350–12,300]	11,038 [8,660–13,810]	12,149 [8,660–14,100]
Urban (10^6 m³)	155 [90–230]	358 [290–430]	630 [570–680]
Environment[a] (10^6 m³)	584	584	584
Total	**10,814**	**11,980**	**13,363**

[a] Excludes winter rice flooding requirements of 123×10^6 m³.

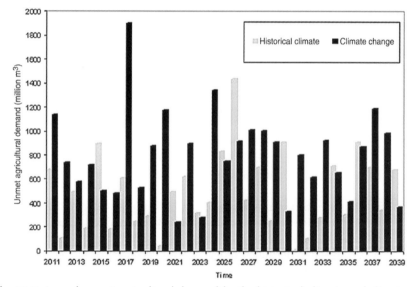

Fig. 11.7. Annual unmet agricultural demand for the historical climate and climate change 2010–2039, with land use changing under both climate regimes.

Upper Pit, Upper Yuba, Upper Feather). The sources of available water for these demands are limited to the rain- and snowmelt-fed tributaries from which they draw. The majority of these unmet demands occur during the summer months of June–September (see Fig. 11.8) – the most critical months in terms of production. The coefficient of variation (defined here as the ratio of the standard deviation to the mean) also increases over these two climate change periods from 0.46 in 2010–2039 to 0.65 in 2070–2099. This has important implications as incomes for farmers may become less certain on an inter-annual basis.

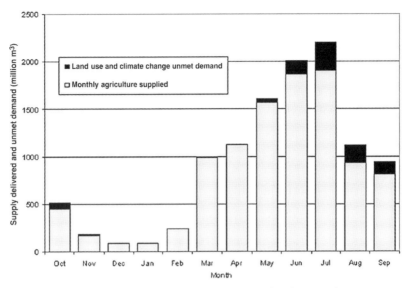

Fig. 11.8. Average monthly agricultural demand supplied and unmet for 2010–2039 (HA2 GCM climate scenario).

Note, as pointed out earlier, the agricultural demands under the climate change scenario are significantly higher due to increased temperatures. The direct impact on the crop water requirements for agriculture are a potential increase in the total water needed of 10% by 2010–2039, and more than 20% by 2070–2099. These changes are validated by the field-scale model, SWAP, for both rice and tomatoes – two of the major crops in the basin. However, the SWAP model also shows increased productivity for these two crops (van Diepen *et al.*, 1989). Rice yields may increase by almost 50% for the A2 and 20% for the B2 scenario, while tomatoes may increase by as much as 20%. This increase in productivity is related to photosynthetically active radiation (PAR), which is used by the plant as energy in the photosynthesis process to convert CO_2 into biomass. Crop production is therefore affected by the air's CO_2 level and in many high-input farming systems the CO_2 levels are the limiting factor in crop production. Important in this process is to make a distinction between C3 and C4 plants. Examples of C3 plants are potato, sugarbeet, wheat, barley, rice, and most trees except mangrove. C4 plants are mainly found in the tropical regions and some examples are millet, maize and sugarcane. The difference between C3 and C4 plants is the way the carbon fixation takes place. C4 plants are more efficient in this, with the loss of carbon during the photorespiration process nearly negligible for C4 plants. Alternatively, C3 plant may lose up to 50% of their recently fixed carbon through photorespiration. This difference has suggested that C3 plants will respond less positively to rising levels of atmospheric CO_2. However, it has been shown that atmospheric CO_2 enrichment can, and does, elicit substantial photosynthetic enhancements in C4 species (Wand *et al.*, 1999).

Demand increases in a period of lower flows with climate change. One would therefore expect more severe levels of unmet demand. However, effects are buffered by groundwater supplies, as discussed in the following section.

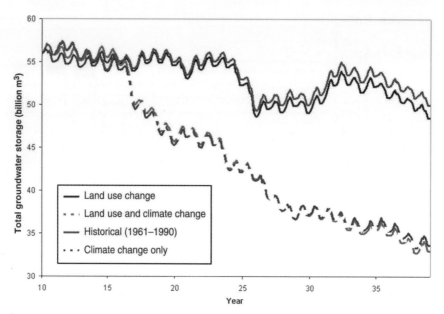

Fig. 11.9. Total groundwater storage using the Hadley A2 scenario 2010–2039 for climate change.

Impacts of climate change: environmental security

As described above, environmental security is measured with four indicators to represent the health of aquatic life and wildlife, water for domestic use and the overall sustainability of the system. Environmental demands for wildlife refuges are consistently met both under conditions of existing climate and climate change. Only in 2070–2099 under the climate change scenario are ecosystem demands unmet, and even then only for on average 1.5% of the time. Average unmet urban demand under existing climate and with climate change is approximately 23×10^6 m^3 and 34×10^6 m^3, respectively, for the period 2010–2039. This represents an unmet demand of 7% without climate change and 10% with climate change in 2010–2039 and 12% in 2070–2099.

An important indicator of environmental security is the change in groundwater storage in the system as a measure of sustainability. There is significant groundwater storage in the Sacramento Basin, which is steadily being depleted in these scenarios, as illustrated in Fig. 11.9. Clearly this pattern of water use is not sustainable. Furthermore, in as much as groundwater usage is driven by agriculture demands and surface water availability, the impacts of climate change alone (without land use changes) are substantial. Under the B2 climate scenario, which in general is more temperate, groundwater storage declines, although not as rapidly as in the A2 climate scenario.

Adaptation Strategies

Historical adaptations to climate variability

Water managers in California have long had to cope with the challenges posed by the state's variable hydrology. The earliest adaptation strategy in response to this variability was to develop simple irrigation systems that allowed for the capture of the base flow available in rivers and streams. The next major adaptation was in response to heavy precipitation and high flow during the 1860s. The extensive flooding during this period led to the creation of the State Reclamation Board, which was charged with the construction of levees to protect growing cities and areas with a high concentration of agricultural production.

With time the population expanded, as did the demand for food. A responsive adaptation strategy was the construction of reservoirs that could carry over winter rainfall and runoff into the summer irrigation season. Protection of growing communities from the risk of flooding also increased in importance. Two additional adaptation strategies were the development of flood control operating rules for large reservoirs and the construction of numerous flood bypasses in the Central Valley.

All of these adaptations, driven by the desire to expand irrigated agriculture in the Central Valley and to reduce the risk of flooding, dramatically altered the hydrology of the Central Valley and the ecosystems that had developed in response to the natural hydrologic regime. Beginning in the 1960s, there began a series of adaptations designed to limit the impact on these important ecosystems. Early adaptations included the establishment of minimum instream flow requirements at important points in the system. More recently the physical rehabilitation of riverine ecosystems has taken on increased importance. Planners now realize that the extensive levies in the Central Valley limit the amount of wetland and riparian habitat available. Levy set-back adaptations are now being considered alongside the concept that flow-bypass structures can be managed as wetland complexes. Already a portion of one bypass in the Central Valley has been converted to a national wildlife refuge. There is also a growing recognition that assuring the proper volume of flow for ecosystems is necessary but not sufficient. Other factors, such as the temperature and quality of the water in rivers, are also important. Recent adaptations with regards to water temperature include the construction of temperature control devices in large dams which allow for the controlled management of cold and warm water pools that generally develop when large reservoirs stratify. Water quality adaptations include the development of discharge permitting requirements. These have been limited to date to point discharges, but are now being contemplated for non-point sources as well.

In summary, the Sacramento Basin has in place a number of adaptations to address natural climate variability in maintaining food and environmental security, as shown in Table 11.3. There has been a clear historical trend towards placing higher priority on environmental security, as people in the basin have come to value the role of ecosystems. Each of the adaptation strategies discussed assumes that several of the existing strategies are essentially maintained, particularly that no reservoirs would be built or destroyed, and there would be no relaxation of water quality constraints.

Table 11.3. Existing adaptation types.

Adaptation type	Food security	Environmental security
Irrigation	X	
Reservoirs	X	X
Levees	X	
Flood bypasses	X	X
Instream flow requirements		X
Conversion of bypasses to wildlife refuges		X
Temperature control devices		X
Water quality permits	X	X

Food security adaptations

The adaptation strategies for food security look at the consequences of reversing the recent policies that prioritized ecosystems to ones that prioritize agricultural production. While urban uses are still given the highest priority, this strategy further imposes demand-side management (DSM) policies that yield a net decrease in urban demands of 20% over the two time horizons. Exports out of the basin, which affect the availability of water to both agricultural and urban areas in southern California, are maintained at present levels.

Environmental security adaptations

The adaptation strategies for environmental security maintain the existing high priority placed on ecosystems, but urban demand side management programmes are implemented (20% reduction by 2100). In addition, the threshold for flooding of the major Yolo bypass is lowered, thus effectively allowing the frequency of diversions of flood waters for wetland habitats to be similar to current conditions. Without this adaptation, the frequency of flooding dramatically drops. Winter rice flooding is adopted throughout the Sacramento Valley to provide additional habitat for migrating birds, requiring an additional demand of 123×10^6 m^3 over the entire winter (December to February). Lastly, restrictions on the maximum withdrawals are made for each aquifer sub-basin. These restrictions are based on the monthly mean abstraction rates during the historical period to prevent the rapid unsustainable decline in groundwater storage that occurs without adaptation in the future.

Integrated adaptations (Water for Food and the Environment)

The adaptation strategies presented in the two previous sections pit the environment against agriculture. However, there are adaptations that can jointly address food and environmental security. One example is the use of groundwater banking, which has received considerable attention in the Sacramento system in recent years as an option

Table 11.4. Summary adaptation strategies.

Adaptation strategy	Measures
1. Land use change	
2. Land use change and climate change	
3. Water for food	Agricultural demands are given priority over the environment Policy of demand-side management leads to a 20% decrease of domestic water use by 2100
4. Water for the environment	Policy of demand-side management leads to a 20% decrease in domestic water use by 2100 Lower threshold for Yolo Bypass flooding to increase wetland, aquatic life and wildlife habitat Adoption of winter rice flooding to increase wildlife habitat Restriction on maximum aquifer withdrawals to ensure sustainability
5. Integration	Groundwater banking node added to the Sacramento Stone Corral agricultural area

for augmentation of critical supplies. A recent study demonstrated that groundwater banking efforts in the Central Valley could potentially provide an additional 1200×10^6 m^3 of annual yield, providing new opportunities for supplying consumptive uses and enhancing stream flows (Purkey, 1998). The basic idea behind groundwater banking is to store excess wet year supplies in subsurface aquifers. Groundwater banking options, unlike the construction of surface water reservoirs, typically are lower cost, less controversial and more efficient, as system-wide losses from evaporation are significantly reduced. Lastly, like other forms of storage, groundwater banking converts fluctuating precipitation and snowmelt into a steady supply stream by storing when water is plentiful and providing when water is scarce.

To explore this, a groundwater bank is added to the system to serve the Sacramento Stone Corral agricultural area (representing approximately 25% of the total agricultural demand in the Sacramento system) and stores excess flows (maximum demand of approximately 1000×10^6 m^3 annually) in the Sacramento River above the Sutter Bypass. The Sacramento Stone Corral demands will first extract needed supplies from the groundwater bank and then resort to the Glenn Colusa and Tehama Colusa canals for additional supplies. Priorities are given such that the Shasta reservoir will release to provide water for the groundwater bank. These adaptations are added to the Water for Environment strategy for comparison. In summary, the adaptation strategies that will be explored in this chapter are given in Table 11.4.

Table 11.5. Average agricultural unmet demand (10^6 m^3) (average % deficit).

Adaptation strategy	2010–2039	2070–2099
Land use change	482 (4.8%)	493 (4.8%)
Land use and climate change	785 (7.1%)	1479 (12.1%)
Water for environment	1052 (9.5%)	1989 (16.3%)
Water for food	719 (6.5%)	1318 (10.8%)

Assessment of Basin-wide Adaptation Strategies

The adaptation strategies described above are measured against the indicators given in the previous section. The effect of the adaptations for each of these indicators is discussed below in more detail.

Food security

Of the three primary water uses affected by land use changes and climate change, agriculture has the largest unmet demands in terms of quantity, but in terms of percentages, the unmet demands are comparable. Agricultural shortfalls, however, are compensated in part by the increased productivity of certain crops due to increased carbon under climate change. As discussed earlier, field-scale modelling using SWAP shows that rice yields, for example, could increase by as much as 50% in the A2 scenario considered here, and tomato yields could increase by 20%. These numbers assume full irrigation.

As discussed in the previous section, with only land use change, agricultural demand will on average have a 5% shortfall in 2010–2039 and 2070–2099. When climate change is introduced into the system, this number increases to 7% over the 2010–2039 period and 12% over the 2070–2099 period (Table 11.5). This analysis further reveals that from a policy perspective, even if the Sacramento system is managed with a focus on food security (Water for Food strategy), about 7% of the demand will still be unmet in 2010–2039, with an increase to about 11% by 2070–2099. When the focus is on environmental security (Water for Environment strategy), these numbers are higher; almost 10% of agriculture demands are on average unmet in 2010–2039, reaching 16% in 2070–2099. Under the Water for Food strategy, unmet demand is slightly less than the land use and climate change scenario because DSM strategies are imposed on all urban demands, thus additional water is available for agriculture uses. For the B2 GCM climate scenarios, the percentage unmet demands are larger. This is in part due to both larger agricultural demands and less available water in the system.

The potentially higher increase in yields, which vary from 20% to 50%, may in fact compensate for these deficits. It is very difficult to predict however, as world prices may react to this increased productivity, which could result in positive or negative impacts on farmer incomes.

In terms of variability of agricultural production (defined as the variation in

Table 11.6. Variation in unmet demand (10^6 m³/year) (coefficient of variation).

Adaptation strategy	2010–2039	2070–2099
Land use change	329 (0.68)	348 (0.70)
Land use and climate change	359 (0.46)	959 (0.65)
Water for environment	524 (0.50)	1204 (0.61)
Water for food	337 (0.47)	802 (0.61)

annual unmet demands), the standard deviation of the unmet agricultural demand increases by almost a factor of 3 under the climate change scenario from 2010–2039 to 2070–2099 (Table 11.6). The coefficient of variation also increases. Furthermore, even with the priority given to agriculture in the Water for Food strategy, the dramatic variation in agricultural production cannot be avoided, although it is slightly reduced. The situation is exacerbated under the Water for Environment strategy, where the standard deviation increases to 524×10^6 m³ in 2010–2039 and increases to 1204×10^6 m³ by 2070–2099.

Environmental security

Throughout the Sacramento Basin a number of instream flow requirements are established through an anadromous fish recovery programme (AFRP). Meeting these requirements is vital not only on the main stem of the Sacramento, but also on tributaries that are critical for spawning. There is also a more general flow requirement downstream to maintain the estuary of the San Francisco Bay – a habitat that provides numerous ecosystem services to the region, including fish, wildlife, water quality, and recreational and aesthetic opportunities.

Four of these instream flow requirements are evaluated for each of the two scenarios and two strategies. Examining the flow requirement on the Feather River, the minimum flow requirements are essentially met under all scenarios and for both time periods. For the remaining three instream flow requirements (shown in Fig. 11.10) and flow requirement on the Sacramento at Freeport for environmental flows to the Delta, it is clear that these requirements are much harder to meet under climate change conditions. Furthermore, the frequency of unmet flow requirements increases in 2070–2099. For instance, for the Delta flow requirement, the frequency of unmet requirements increases by almost a factor 2. Similar to the situation for wetlands, the Water for Food strategy is more problematic for each of the instream flow requirements. Lastly, examining the American AFRP (Fig. 11.10), we observe that typically in the months of May–July the flow requirements are unmet.

Wetlands

Under the two adaptation strategies, unmet demands related to wetland areas on average increase over time due to climate change. For 2010–2039 under the Water for Environment strategy, unmet demands are 1.7%. However, this is largely due to additional environmental demands introduced (i.e. winter rice flooding) and the

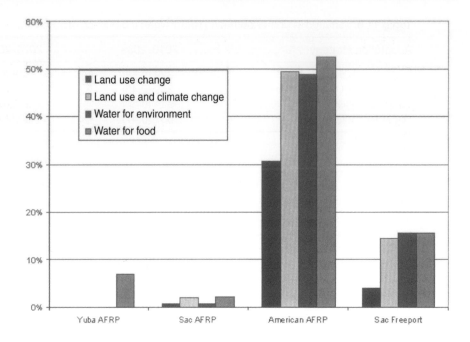

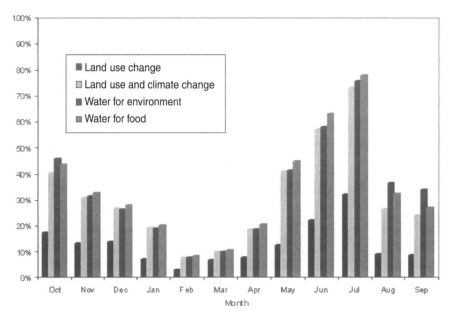

Fig. 11.10. Frequency of unmet monthly flow requirements 2010–2039 (top) and average percentage unmet flow requirement 2070–2099 for the American AFRP (bottom).

Table 11.7. Frequency of annual flow events to the Yolo bypass relative to current conditions.

Adaptation strategy	2012–2039	2072–2099
Land use change	46%	41%
Land use and climate change	29%	32%
Water for environment	39%	32%
Water for food	29%	32%

restrictions on groundwater withdrawals imposed. Thus, effectively, despite these small unmet demands, total wetland area increases under this strategy. Under the Water for Food strategy, environmental unmet demands increase to 4.5%. This percentage increases to 5.1% under the 2070–2099 scenario. These demands are generally unmet during the autumn/winter months when the wetland and refuge requirements are the highest, thus the relative impact of these shortfalls are greater than indicated by the annual average percentage (Table 11.7).

Domestic water-supply sector

Urban demands are given the highest priority in the Water for Food strategy. As a result, the percentage unmet demands are the lowest for this strategy. Unmet demands in the Water for Environment strategy are due in large to the priority given to American AFRP requirements downstream of Placer and El Dorado counties. Unmet urban demands are comparable in terms of percentage to agricultural unmet demands and increase over time in the presence of climate change. Demand management is implemented as an adaptation to partly mitigate losses taken in the Environment and Food Security strategies, thus dampening the overall loss of water for this sector. Furthermore, under the B2 climate scenario, unmet urban demands increase because of less available water in the Sacramento system as a whole (Table 11.8).

Groundwater storage

Groundwater storage in general declines across all strategies. However, the rate of decline varies. For the Water for Environment strategy, the imposed groundwater withdrawal restrictions limit the decline in storage in both the 2010–2039 and 2070–2099 scenario. Storage still declines under this strategy at a rate of about 275 million m^3/year, a factor of almost 3 less than in the Water for Food strategy (860 million m^3/year). The effects of land use change and climate change on storage are also evident. More groundwater is extracted under the combined land use and climate change scenario because of increased evaporative demands for agriculture. Furthermore, under the Water for Food strategy, current rates of withdrawal are clearly unsustainable. These findings can be generalized across individual groundwater sub-basins, although the severity of depletion varies widely. For instance, in the Sacramento Stone Corral groundwater sub-basin, under the Water for Food and Land Use and Climate Change scenarios, the aquifer is near depletion by 2050.

Table 11.8. Average urban unmet demand (average % deficit).

Adaptation strategy	2010–2039	2070–2099
Land use change	6.5%	6.3%
Land use and climate change	9.5%	11.7%
Water for environment	8.3%	22.0%
Water for food	0.0%	10.4%

Integration (Water for Food and the Environment)

For the 2010–2039 time period, the total unmet agricultural demand decreases from 9.5% (Water for Environment) to 7.8%. Similarly, by 2070–2099, the total unmet agricultural demand decreases from 16.3% to 13.8%. This approximate 2% improvement can directly be attributed to the improved release at the Sacramento Stone Corral. In the Water for Environment strategy for 2010–2039, the unmet demand at the Sacramento Stone Corral is on average 6%. With groundwater banking, unmet demands are negligible. Furthermore, the inter-annual variability in unmet demands also declines from $524 \times 10^6 \, m^3$ to $425 \times 10^6 \, m^3$ (~20% reduction). Unmet demands for urban and environmental demands are unaffected as these are given higher priorities in the Water for Environment strategy. The one drawback to such a strategy is that the flows through the Yolo Bypass are reduced by almost 10% on an annual basis. A more sophisticated analysis would be required to determine the trade-offs of the gains from agricultural production and instream habitat versus wetland impacts in the Yolo Bypass.

Clearly, the use of groundwater banking can provide the Sacramento system with a win–win situation, as is illustrated by the example at Sacramento Stone Corral. Such a programme can be implemented in most agricultural areas in the Central Valley. By better managing groundwater aquifers, the overall supply of water available to the entire system can be increased, thus reducing unmet demands. Furthermore, on the environmental side, by banking groundwater the rapid declines in groundwater storage can be slowed (Fig. 11.11), making it easier to achieve groundwater sustainability goals.

Summary and Conclusions

Climate change clearly has serious implications for water management in the Sacramento Basin. Based on the results presented above, several conclusions can be drawn which are summarized in Tables 11.9 and 11.10. The Land Use Change scenario (i.e. no climate change) is taken as a reference or business as usual scenario for comparing climate change impacts and adaptation strategies.

First, in terms of agricultural production, it is clear that a certain level of unmet agricultural water demand is unavoidable and that climate change will exacerbate the situation. Unmet agricultural demands increase by about 2.3% from the business as usual scenario when climate change is introduced. This increased unmet demand, however, may be offset by the increased yields of rice (possible increase of up to 50%) and tomato crops (possible increase of up to 20–30%). For instance, given that rice

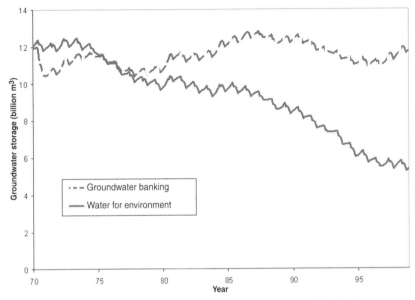

Fig. 11.11. Groundwater storage at the Sacramento Stone Corral for both the water for environment and integration strategies and Hadley A2 GCM scenario for 2070–2099.

accounts for a modelled agricultural area of 225,000 ha (19% of the total irrigated area in the Sacramento Basin) and assuming that the total unmet demands are distributed proportional to the irrigation requirements and that acreage is a surrogate for production, these yield increases could balance the 2.3% unmet demands under climate change. A key question for further research is to determine the extent to which these increases in water shortfalls (an additional 2.3% in 2010–2039 and 7.4% in 2070–2099) can be offset by increased productivity, stabilizing the overall production and income generated for farmers.

By adopting a policy of food security, unmet agricultural water demand (above the reference) decreases from 2.3 to 1.7% for 2010–2039, a relatively small impact. Similarly, for 2070–2099, the Water for Food strategy shows a reduction in the amount of unmet demand by a little over 1%. However, what is gained by such a strategy in both periods is a reduction in the variability in these unmet demands. Thus, such strategies are important mechanisms for effectively reducing the inter-annual risk farmers face in terms of agricultural production. Furthermore, in both time periods, these increases in unmet demands can be offset by the use of groundwater banking, as illustrated in the integration (Water for Food and the Environment) strategy. For the Sacramento Stone Corral, for instance, groundwater banking means that on average all demands are met.

One possible adaptation strategy that was not considered here is the reduction in total agricultural area by switching to higher value crops. Such a policy initiative may make an appreciable difference. Also, there is a ticking time-bomb in the overall water use – in the business as usual and Water for Food strategies, groundwater is being unsustainably mined, and even calls into question whether agriculture would continue to be profitable given the possible pumping costs associated with this mining.

Table 11.9. Effect of adaptations on food and environmental security 2010–2039.

Adaptations	Agricultural production (1)	Variance in agricultural production (2)	Salmon population (3)	Wetland area (4)	Domestic water supply (5)	Ground-water storage
Land use and climate change	−2%	1.1X increase	−6%	No change	−3%	Fast decline
Water for food	−2%	No increase	−8%	−5%	No change	Fast decline
Water for environment	−5%	1.6X increase	−6%	Slight increase*	−2%	Slow decline
Integration	−3%	1.3X increase	−6%	Slight increase*	−2%	No decline

(1) Average percentage change in total met agricultural demand from a baseline of 5% unmet demand.

(2) Factor increases above the baseline standard deviation of 328×10^6 m^3.

(3) Percentage change in average met flow requirements. Based on an average of four AFRPs (Yuba River, American River, Feather River and Sacramento River) and Freeport flow requirement.

(4) Average percentage change in total met environmental demand (refuges and wetlands) from the baseline of 0% unmet demand. *The environmental demands for both the Water for Environment and Integration strategies increase by 123×10^6 m^3. Thus, although 2% of these increased demands are unmet, compared to the baseline, effectively there is an increase in total wetland area.

(5) Average percentage change in total met urban demand from the baseline of 6.5% unmet demand.

Groundwater banking provides a solution to this critical issue while improving the availability of supplies during dry season events at the same time.

Groundwater storage will decline as the aquifers are being pumped at rates beyond annual renewable recharge unless some intervention is made. For example, by imposing constraints on the maximum amounts that can be withdrawn from the aquifers in the Water for Environment strategy, the rate of decline is significantly slowed. In the Integrated strategy, groundwater banking is shown to further slow, if not eliminate the rate of decline.

For the environment, it is shown that certain flow requirements related to aquatic ecosystem health (AFRPs) will be difficult to meet (e.g. American River AFRP), while others are relatively easily met (e.g. Feather River AFRP). Strategies that prioritize these flow requirements result in a marginal improvement compared to strategies that prioritize municipal and agricultural demands. With growing stresses on water availability, unmet demands for wetlands and refuges may be as much as 5%, unless the environment is prioritized. These levels of unmet demands may or may not be acceptable depending both on the timing of the deficits (e.g. during critical spawning

Table 11.10. Effect of adaptations on food and environmental security 2070–2099.

| | Indicators | | | | | |
	Agricultural production (1)	Variance in agricultural production (2)	Salmon population (3)	Wetland area (4)	Domestic water supply (5)	Ground-water storage
Adaptations						
Land use and climate change	−7%	2.8X increase	−14%	−1.5%	−5%	Fast decline
Water for food	−6%	2.3X increase	−19%	−5%	−4%	Fast decline
Water for environment	−12%	3.4X increase	−12%	Slight increase*	−12%	Slow decline
Integration	−9%	3.4X increase	−12%	Slight increase*	−12%	No decline

(1) Average percentage change in total met agricultural demand from the baseline of 5% unmet demand.
(2) Factor increases above the baseline standard deviation of 348×10^6 m^3.
(3) Percentage change in average met flow requirements. Based on an average of four AFRPs (Yuba River, American River, Feather River and Sacramento River) and Freeport flow requirement.
(4) Average percentage change in total met environmental demand (refuges and wetlands) from the baseline of 0% unmet demand. *The environmental demands for both the Water for Environment and Integration strategies increase by 123×10^6 m^3. Thus, although 2% of these increased demands are unmet, compared to the baseline, effectively there is an increase in total wetland area.
(5) Average percentage change in total met urban demand from the baseline of 6.5% unmet demand.

periods) and on the social value assigned to the ecosystems. By adopting a Water for Environment strategy or Integrated strategy, wetland areas can be increased. Under these strategies, only approximately 2% of these increased demands are unmet.

The business as usual scenario will result in about 6% unmet urban demand. This is primarily due to counties that draw municipal water from the American River upstream of higher priority flow requirements. Whether or not the flow requirement would be prioritized over municipal demands in the future is unknown, but poses an interesting trade-off question. Unmet urban demands are, in general, higher when the environment is prioritized and lower when food security is prioritized (it is assumed that urban demands would always be satisfied before agriculture, although subject to demand management policies). For instance, by 2070–2099, unmet demands are estimated to increase by 5% above the business as usual case in the presence of climate change. By adopting a Water for Food strategy, unmet urban demands only increase by 4%. But by adopting a Water for Environment strategy, unmet demands increase by 12%.

These results are summarized in Tables 11.9 and 11.10.

The adaptation strategies explored here generally involve making trade-offs between food and environmental security, as is highlighted by Tables 11.9 and 11.10. Under Water For Food, agricultural security improves at the cost of the environment. Similarly, under Water for Environment, shifting water to the environment entails some shift of water out of agriculture. Moreover, the situation worsens in 2070–2099 as climate change effects become more pronounced. Mitigations of trade-offs come into play through the enhancement of productivity of crops due to increased carbon, finding non-water consuming activities to support the economy of the region, and implementation of integration win–win strategies such as groundwater banking.

References

Department of Water Resources (1998) The California Water Plan Update Bulletin, 160-98. Department of Water Resources, Sacramento, California. Available at: http://rubicon.water.ca.gov/pdfs/ b160cont.html#es

Gutowski, W.J., Pan, Z., Anderson, C.J., Arritt, R.W., Otieno, F., Takle, E.S., Christensen, J.H. and Christensen, O.B. (2000) *What RCM Data are Available for California Impacts Modeling?* California Energy Commission Workshop on Climate Change Strategies for California, 12–13 June, California Energy Commission, Sacramento, California.

IPCC (2001) Climate Change 2001. The Scientific Basis. Contribution of Working Group I to the Third Assessment Report of the Intergovernmental Panel on Climate Change. Available at: http://www.grida.no/climate/ ipcc_tar/wg1/index.htm

Purkey, D. (1998) Feasibility Study of a Maximal Program of Groundwater Banking, Natural Heritage Institute Report, December 1998. Available at: http://www.n-h-i.org/ Publications/Pubs_pdf/ Conj_use.pdf

van Diepen, C.A., Wolf, J., van Keulen, H. and Rappoldt, C. (1989) WOFOST: a simulation model of crop production. *Soil Use and Management* 5, 16–25.

WEAP (2002) Water Evaluation and Planning System. Available at: http://www.seib.org/ weap

12 Food Demand and Production: a Global and Regional Perspective

KENNETH STRZEPEK,[1] ALYSSA MCCLUSKEY,[1] JIPPE HOOGEVEEN[2]
AND JOS VAN DAM[3]

[1]University of Colorado, Boulder, Colorado, USA; [2]FAO, Rome, Italy;
[3]Wageningen University, Wageningen, The Netherlands

Driving Forces

The most important driving force for future agricultural production is population growth. The current population of about 6 billion people will grow to more than 8 billion in 2030. The world's population is still growing at an impressive rate, but slower than before. This situation translates into an overall reduction of the growth rate of agricultural production, although in absolute terms a massive increase of production will still take place between now and 2030 to satisfy the growing demand. For example, Seckler *et al.* (1999) estimated that by 2025 cereal production will have to increase by 38% to meet world food demands. The World Water Vision, an outcome from the Second World Water Forum in The Hague in 2000, estimated a similar figure of 40% based on various projections and modelling exercises (Cosgrove and Rijsberman, 2000). Adopting UN mid-range population estimates of 8.9 billion people combined with the minimum caloric requirement of 2200 calories per day, means that a total of about 20 trillion consumable calories have to be produced. Current levels are about 14 trillion calories, which means an increase of 42%. However, given the range in population estimates as provided by the UN this figure can be between 14% and 71%.

The demand does not grow only because of an increasing number of people, but also because they eat more. When GDP per capita rises, the food preference of the people changes and the daily calorie intake per capita rises. Overall crop demand and production will therefore increase considerably more than the increase of the population. Increase in crop demand can be met in three different ways: expanding the land area, increasing the frequency with which it is cropped and boosting yields (Fischer *et al.*, 2003). It has been suggested that ceilings may be approached in all three factors, but FAO's study, *World Agriculture: Towards 2015/2030*, does not support this view at the global level (FAO, 2002). About 80% of the projected growth in crop production in developing countries comes from yield increases (67%) and intensification of the cropping pattern (12%). The rest will come from extending cultivated land.

RESULTS   **MODELS**   **DRIVERS**

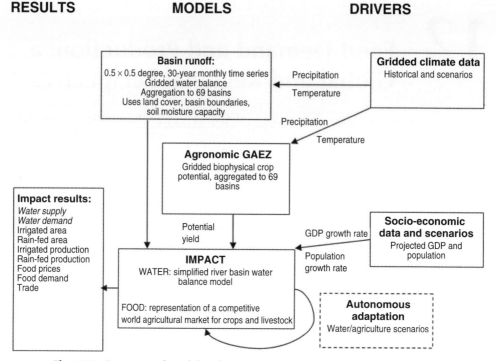

Fig. 12.1. Overview of model and data components related to the IMPACT model.

The estimated contribution of yield increases and intensification of the cropping pattern is partly the result of the increasing share of irrigated agriculture in total crop production, since the yields of irrigated crops are generally higher and irrigated agriculture is generally more intensive than rain-fed agriculture.

This chapter deals with the results of a study that aimed at assessing global impacts on food demand and security under climate change and autonomous adaptation. The core of the study is an integrated modelling framework that allows assessment of food trade, food demand and food production, the latter coupled to a global water balance model. The results are compared with the effects at basin and field scale as discussed in Chapter 3.

A Global Assessment Study

Figure 12.1 summarizes the overall approach used in the integrated global analysis of the impacts of climate change and climate variability on the availability and use of water and the production and consumption of food. At the core of the approach is the IMPACT model, which is a representation of a competitive world agricultural market for crops and livestock. It is specified as a set of country or regional submodels, within each of which supply, demand and prices for agricultural commodities are determined. The country and regional agricultural submodels are linked through trade, a specification that highlights the interdependence of countries and commodities in the global agricultural markets. World agricultural commodity prices are determined annually at levels that clear international markets.

In this study, IMPACT has been supported by two additional models that provided forcing data to drive it. The first model is a global hydrology model 'IMPACT-WATER' that uses first-order data (climate, land cover, soil type, etc.) specified at a $0.5° \times 0.5°$ spatial grid for all global land points (excluding the Arctic and Antarctica) to produce estimates of river basin runoff over a 30-year time horizon and on a monthly time step. In the IMPACT-WATER model, water is represented as a scarce resource needed for agricultural production. Water demands evaluated in the water model include irrigation, livestock, domestic, industrial and committed flow for environmental, ecological and navigational uses. The river basin runoff data are used by ADAPT to define the water supply to each of the hydro-economic zones. A second external model, the Global Agro-Ecological Zones (GAEZ) model is used to determine crop potential yield. The GAEZ provides a standardized framework for the characterization of climate, soil and terrain conditions relevant to agricultural production, most notably the estimate of maximum potential crop yield in a gridded format that can be used by IMPACT. In GAEZ, crop modelling and environmental matching procedures are used to identify crop-specific limitations due to climate, soil and terrain, under assumed levels of inputs and management conditions. The GAEZ model was derived from Fischer *et al.* (2003).

Since a keen interest of this study is adaptation, the lower right corner of Fig. 12.1 expresses the implicit adaptation that can be accounted for in the IMPACT model. IMPACT is a representation of a competitive world agricultural market for tradable crops and livestock, which determines supply, demand and prices for these commodities and determines their price such that international markets clear. Thus, different climate or socio-economic scenarios trigger autonomous adaptation implicit in the IMPACT model results. Autonomous adaptation implies either gradual or abrupt changes in food production, such as increases in acreage, changes in water use, changes in cropping patterns over time, that are result of the dynamics of a competitive food market, constrained by production factors such as a limited water supply or limited agricultural capital (labour, land, mechanization, etc.).

Although the IMPACT model will account for autonomous adaptation, IMPACT is also capable of describing the response of global water use and food production to exogenous adaptation strategies such as improved irrigation efficiency, changes in farm subsidies, capital investments, etc. These external or exogenous variables would be prescribed, and would comprise individual scenarios whose results would be examined relative to baseline, non-exogenously driven scenarios. For this study, we have focused on autonomous adaptations, with the hopes of extending the work to include scenarios of explicit, exogenous adaptation strategies in the future.

The IMPACT model divides the world into 69 broad hydro-economic regions, with three of these regions (the USA, China and India) divided more finely into their major river basins. For each basin, the mean monthly runoff derived for the period 1961–1990 was compared with an estimate of monthly mean runoff from Alcamo *et al.* (2000), which was used to calibrate the global water balance used in this study (Yates, 1996). The runoff model was calibrated by comparing the 'observed' monthly average discharge from the Alcamo dataset, to model estimates of runoff, where

Table 12.1. HadCM3 and ECHAM4 global temperature projections (A1 and B2).

	A1 Global mean temperatures relative to 1990 (°C)										
	2000	2010	2020	2030	2040	2050	2060	2070	2080	2090	2100
HadCM3	0.570	0.730	0.989	1.385	1.863	2.253	2.695	3.081	3.383	3.621	3.823
ECHAM4	0.379	0.488	0.671	0.962	1.315	1.595	1.923	2.209	2.434	2.614	2.770

	B2 Global mean temperatures relative to 1990 (°C)										
	2000	2010	2020	2030	2040	2050	2060	2070	2080	2090	2100
HadCM3	0.570	0.839	1.163	1.468	1.768	2.070	2.362	2.653	2.943	3.229	3.510
ECHAM4	0.379	0.577	0.813	1.033	1.251	1.472	1.686	1.900	2.115	2.328	2.538

monthly runoff is simply the accumulation of the surface and subsurface components. Calibration of the regions and basins that comprise the 69 hydro-economic zones was done by enumerating values that would minimize the RMSE error between the 'observed' mean monthly discharge, and the estimated discharge from the model for each of the 69 basins. Thus calibrations for each basin result in a unique correlation value for each land cover within that basin.

Global GCM-based Scenarios

The IPCC published a Special Report on Emissions Scenarios (SRES) in 2000 (New *et al.*, 2000; see Chapter 2). SRES describes a new set of emissions scenarios used in the Third Assessment Report. The SRES scenarios have been constructed to explore future developments in the global environment with special reference to the production of greenhouse gases and aerosol precursor emissions. The SRES scenarios were used as atmospheric forcing functions for a suite of Global Circulation Models (GCMs) that provide simulations of future climatic variables. The GCMs used in this analysis to provide precipitation and temperature for climate change scenarios include Max Planck Institute Model ECHAM4 for B2 2020, B2 2080, Hadley Center Model HadCM3 for A1 2020, A1 2080, B2 2020, and B2 2080. Table 12.1 shows the global mean temperatures for both GCM outputs for the years 2000–2100.

ENSO variability scenarios

In addition to GCM-based climate scenarios, a statistical technique was developed to derive new scenarios based on changes in the occurrence of the El Niño Southern Oscillation phenomenon (ENSO). The ENSO are sea-surface temperature anomalies that occur periodically off the western coast of South America, with a seasonal timescale. When sea-surface temperatures are above normal, the condition is referred to as El Niño, while if the sea-surface temperatures are below normal, the condition is called La Niña. Table 12.2 shows the years of El Niño and La Niña over the period

Table 12.2. The ENSO years, characterized by either a La Niña or El Niño condition. All other years (68 in total) were considered 'neutral' years.

La Niña	El Niño
1909	1903
1910	1906
1917	1912
1918	1915
1925	1919
1943	1926
1950	1931
1956	1941
1971	1942
1974	1958
1976	1966
1989	1973
	1983
	1987
	1992
	1995

1901–1996. In all, 28 years were considered to be either El Niño or La Niña, while the remainder were considered 'neutral' years.

Socio-economic data and other scenarios

Extensive socio-economic data are required for the IMPACT-WATER modelling framework. The information is drawn from highly disparate databases and requires an interdisciplinary and international collaboration of professionals in agronomy, economics, engineering and public policy. Table 12.3 describes the major data and their sources, which are classified into six classes: water supply infrastructure, hydrology, agronomy, crop production and non-irrigation water demand and water policies. The data have been prepared for river basins (in China, India and the USA) and countries and regions. Data are calibrated for the base year and are then determined by the model for future years (including irrigated and rain-fed crop area and yield, and crop area and yield reduction from water shortages). As indicated above and in Table 12.3, some data came directly from other sources, some are treated based on other sources, and some are estimated from related literature.

GIS and other methods are used to treat these parameters. Some data are given in spatial units smaller than the hydrologic units (such as for China, the USA and districts in India), and the GIS program is applied to overlay the data at the smaller scales. Many other intermediate programs were developed to estimate the required data or transfer the original data to the format required by the models. Data required for agricultural modelling by IMPACT are described in Rosegrant *et al.* (2001).

Table 12.3. Major sources of data used as input for the IMPACT-WATER model.

Category	Details	Sources
Infrastructure	Reservoir storage Withdrawal capacity Groundwater pumping capacity Water distribution, use and recycling situation	ICOLD (1998) WRI (2000); Gleick (1993) WRI (2000) Scenario Development Panel, World Water Vision
Hydrology	Watershed delineation Precipitation Potential evapotranspiration Runoff Groundwater recharge Committed flow Water pollution (salinity)	WRI Basin Runoff Model Basin Runoff Model Basin Runoff Model WRI (2000); Gleick (1999) Authors' assessments Authors' assessments
Agronomy	Crop growth stages Crop evapotranspiration coefficients (k_c) Yield–water response coefficient (k_y)	Rice provided by FAO; wheat and maize by CIMMYT; and other crops by USDA FAO (1998); Doorenbos and Kassam (1977) FAO (1998); Doorenbos and Pruitt (1979)
Crop production	Irrigated and rainfed area (baseline): actual harvested and potential Irrigated and rainfed yield (baseline): actual and potential	FAO (1999); Cai (1999) GAEZ model
Non-irrigation water demand	Industry Domestic Livestock	Shiklomanov (1999) for the Scenario Development Panel, World Water Vision Shiklomanov (1999) for the Scenario Development Panel, World Water Vision Mancl (1994); Beckett and Oltjen (1993); FAO (1986)
Water policies	Committed flows Water demand growth International water sharing agreements Investment	Authors' assessments Authors' assessments Authors' assessments based on WRI (2000) Authors' assessments

Source: Compiled by authors.
CIMMYT, the International Wheat and Maize Improvement Center; FAO, the Food and Agriculture Organization of the United Nations; ICOLD, International Commission on Large Dams; WRI, World Resources Institute; USDA, the United States Department of Agriculture.

Adaptation at global scale

This study addresses two types of spontaneous adaptation to climate change: agronomic and economic. These adaptation types will be manifested in irrigated and rainfed areas, production, yields and producer prices as well as consumer food demand and international trade and resulting world market prices of crop commodities. The global adaptations can be briefly described as:

1. *Agronomic:* The GAEZ analysis for climate change adjusts the planting dates of each crop in each grid cell to maximize the potential yield based on the temperature regime from each GCM scenario.

2. *Economic:* With the change in precipitation, consumptive use and water supply due to climate change, the 'crop per drop' will change. In economic terms, the marginal value of water to irrigated crops will change. The food model will then allocate the available water to the crops based on the higher marginal return amongst the crops, including the impacts on rain-fed yields. The marginal value is determined not only by the supply side but also by the demand for these crops. The micro-economic assumption of market clearing leads to the model 'adapting' the crop production, imports and exports to balance food demand at each region and globally. This type of market adaptation to climate change is known as 'autonomous' adaptation.

Global Results

In 1995 global freshwater withdrawals were approximately 15% of global runoff. The geographical distribution of supply and demand leads to regions of excess and regions of scarcity. Thus any 'global' analysis of climate change impacts on the water system are misleading, as the global mean temperature increases are averages of regions with significantly more and significantly less warming. Any single 'global' indicator of climate change impacts on water resources is a very crude measure of the impacts. In this section we will report on a series of hydro-climatic parameters calculated by the models. These are growing season, potential evapotranspiration and effective precipitation, crop water deficit and available irrigation water supply. These variables are presented for the globe as a very crude measure. To provide some further insight and geographic resolution, these variables are presented separately for the developed and developing regions of the world.

Figure 12.2 shows that globally there is very little change from the base conditions for actual crop evapotranspiration and water available for irrigation. However, globally there is a significant decrease in effective precipitation and thus an increase in water deficit. These results show that rain-fed agriculture will be impacted more than irrigated agriculture.

Results were aggregated into developed and developing regions. The results show that the developing regions are more strongly impacted. There is a greater decrease in effective precipitation and a greater increase in water deficit which is not made by the small increase of available water.

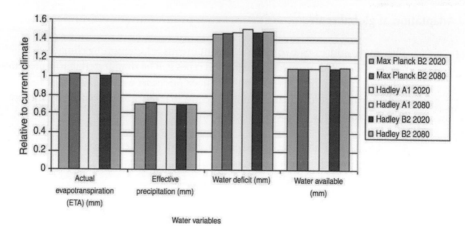

Fig. 12.2. Global water analysis of climate change in the 2080s relative to current climate.

Mean value climate change

The IMPACT-WATER model simulated the time period 1996–2025 with climate change acting as shock to the climate. We are reporting the results from 2000 onward to allow for model adjustments to the 'mathematical shock'. The model is driven not only by climate, but also by socio-economic change and the associated technological changes. As seen in Fig. 12.3, the prices of wheat decline from 2000 to 2025 as the economic impacts of yield increase are felt by the system. Note, however, that the climate change scenarios preserve this trend, but impact the level of final prices. This same phenomenon is illustrated in Fig. 12.4, showing an increase in global food demand mainly as the result of increased population and lower prices. Climate change again impacts the rate and level of this increased demand.

   To present as much information as possible from the analysis, the impact of climate change on food price, demand and production will be presented as changes in the average of the variable over the period 2001–2025. Please note that what is being presented is the *relative* change from the base, not climate change condition. So while demand is presented as decreasing, that is only in respect to the base; in fact it has been increasing in an absolute sense even under climate change.

   A series of graphs present the results for three main crops: wheat, rice and maize. Figure 12.5 shows that for all climate change scenarios and all three crops, total production goes down. Examining the above figures shows that rain-fed production of all crops decreases significantly due to decreases in effective precipitation. Irrigated production decreases more for wheat and rice, but increases for maize. The decrease in global production results in an increase in world market prices. Rice shows a price increase over 100% and wheat almost 100%, while maize experiences a less than 50% increase. These price changes correlate directly to the production changes (Fig. 12.6).

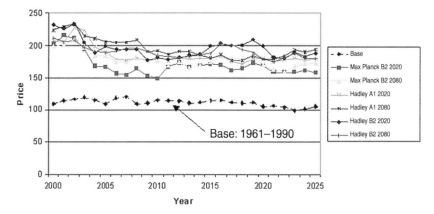

Fig. 12.3. World market prices for wheat.

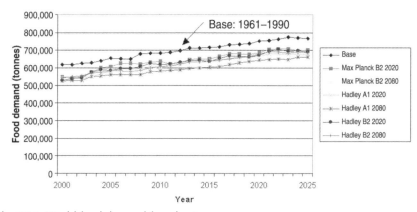

Fig. 12.4. World food demand for wheat.

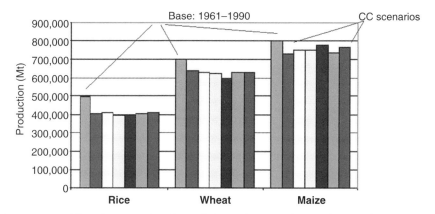

Fig. 12.5. World total (irrigated and rain-fed) production of rice, wheat and maize.

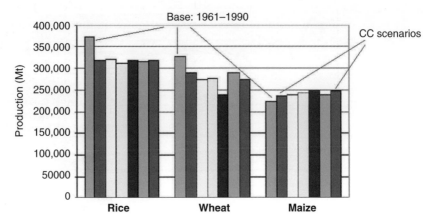

Fig. 12.6. World irrigated production of rice, wheat and maize.

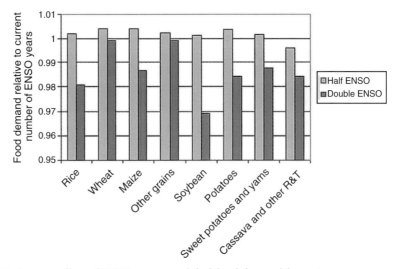

Fig. 12.7. Average effect of ENSO years on global food demand for major crops.

Climate variability

GCM-based climate scenarios only provide estimates of mean monthly changes. How do changes in climate variability (such as an ENSO event) affect food production, prices and demand? Two climate variability scenarios were developed and run through the IMPACT-WATER model and results are presented below. Figure 12.7 shows the impact of increased and decreased ENSO events on global food demand. While halving ENSO has little effect, doubling ENSOs has appreciable effects. Finally, the impact of the variability scenarios on a range of model variables shows that developing regions are impacted more than developed regions.

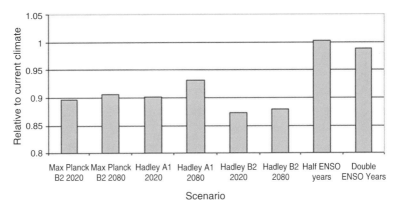

Fig. 12.8. Effects of climate change and variability on global food demand.

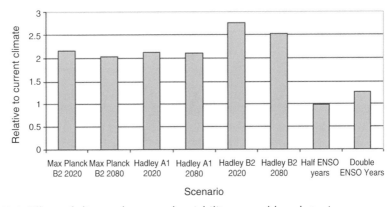

Fig. 12.9. Effects of climate change and variability on world market prices.

Climate change versus climate variability

A question often posed is: what is the relative importance of change in mean versus changes in the variability of climate variables on food systems? Since this assessment examined both type of scenarios we are able to present some comparative results. Figure 12.8 shows the relative impact on global food demand and Fig. 12.9 on food prices for climate change versus climate variability. The results show that for food demand the climate change scenarios have a much more significant impact on food prices and thus food demand.

Global Results Versus Regional Projections

In Table 12.4 a summary is presented with the increases in yield and production in cereals for the year 2030. These are projections according to: (i) field-scale modelling presented in Chapter 3; (ii) projections by FAO; and (iii) model simulations by IMPACT.

Table 12.4. Changes in cereal yields and production towards 2030.

	Field scale		FAO	Global impact
	Average cereal yield in 1998 (t/ha)	Increase in total yield in 2030 (%)	Increase in total production in 2030 – FAO (%)	Increase in total production in 2025 – IMPACT (%)
Volta	0.7	72	152	195
Zayandeh Rud	4.5	19	66	41
Walawe	3.5	56	47	83
Mekong	2.7	43	80	111
Syr Darya	2.3	36	78	25
Rhine	5.5	34	24	15
Sacramento	5.6	12	25	44

Ad (1)

Droogers and van Dam (2004) applied the SWAP (Soil–Water–Atmosphere–Plant) model to analyse the effects of climate change, including CO_2 enhancement, on the crop yields for the seven ADAPT catchments (see Chapter 3). The main conclusion of their study is that in almost all catchments negative effects of prolonged droughts or extreme wet periods are offset by the positive effects of higher CO_2 concentrations. Due to climate change, average crop yields will increase by about 20% in the period 2010–2039 and by about 30% in the period 2070–2099. These yield increases are not yet included in the analysis by FAO and the global analysis, which means that their production levels are on the safe side.

Ad (2)

For the ADAPT project, crop production tables have been prepared for 1998, 2030 and 2050 based on official FAO projections (see, for example, Table 12.4). These tables have been prepared for all basins and contain information for all major crops in the basins on harvested land, crop yield and crop production. For the basins in developing countries a distinction has been made between agricultural production coming from irrigated and rain-fed land. For the Rhine and the Sacramento Basins, this distinction could not be made.

Ad (3)

One of the constraints in the global IMPACT analysis is the restriction of specifying basin boundaries. The IMPACT models are currently hardwired for a set of 69 basins, referred to as the IMPACT basins/regions. The seven ADAPT case study basins were mapped to the IMPACT basins/regions. The basin representation is more acceptable in regions that are fairly homogeneous and small. Some basins were split over more than one region, for example the Volta was split between Northern Africa and Central and Western Africa.

Table 12.5. Global IMPACT results: climate change impacts on cereal production in 2025 relative to current situation.

	Cereal production in 2025 relative to 1995 (%)						
Climate scenario	Sacramento	Rhine	Syr Darya	Volta	Zayandeh Rud	Walawe	Mekong
Current climate	144	115	125	295	141	183	211
MaxPl B2 2020	170	133	126	268	111	181	204
MaxPl B2 2080	178	135	119	259	101	181	200
HadA1 2020	184	142	130	264	113	182	198
HadA1 2080	178	144	108	258	89	176	189
HadB2 2020	186	136	122	274	116	181	205
HadB2 2080	184	142	119	267	107	182	202

In the FAO (2002) study, for the base year (1998) the total crop yields for cereals vary from 0.7 t/ha in the Volta to 5.6 t/ha in the USA. In the Volta Basin, the major cultivated cereal is sorghum under rain-fed conditions with an average yield of only 0.8 t/ha. The only crop that is cultivated under irrigation is rice, with an average yield of 2.4 t/ha. In the Sacramento Basin, rice is also being cultivated, but with average yields of 6.5 t/ha. In the case of the USA, maize is the cereal that is responsible for the high average cereal yields, with a yield of 8.3 t/ha, which is very high if compared with the 1.3 ton/ha that is obtained in the Volta Basin.

From Tables 12.4 and 12.5 it can be concluded that total production generally exceeds the increase in yields. This means that more land will be harvested (either due to expansion of agricultural land or due to an increase in cropping intensity). Only in the case of the Rhine Basin, the yield increase is higher than the increase in production, which implies that agricultural land will be taken out of production. This is not surprising, since the pressure of other sectors on agriculture in this part of the world is very high and the population is practically not growing anymore.

Table 12.5 further shows that in all basins an increase in yield is expected. The highest percentage increase is expected in the Volta Basin (72%) and the lowest in the USA (12%). Climate change has not been taken into account explicitly in these expected yield increases from the FAO (2002). They expected that the overall effects of climate-induced changes in crop production factors and yields are small compared to those stemming from economic and technological growth.

At the latitudes where most of the ADAPT basins are situated, changes in temperature and precipitation are generally expected to have a negative impact on crop yields. However, the rise in atmospheric concentration of CO_2 can also be a positive factor in crop production. It stimulates photosynthesis and it improves the water use efficiency of the crop. Over the last decades an enormous amount of research has been done on the impact of changes in CO_2 concentrations on crop growth. An extensive review of these experiments has been provided by Jones and Curtis (2000) and the Center for the Study of Carbon Dioxide and Global Change (CSCDGC, 2002). They show that in case of doubled CO_2 concentrations under water-limiting conditions, crop production increases by 35–40% for C3 crops, and by about 25% for C4 crops.

The far right column of Table 12.4 lists the changes in cereal production for 2025 as projected by IFPRI/IMWI's IMPACT model used in the global analysis of the ADAPT project. The values for IFPRI/IWMI are for entire economic regions in which the basins fall and do not correspond exactly with the regions used by the FAO. However, the general trends of the estimates are similar, with an R^2 of 0.7 for a linear fit of the two data sets.

Table 12.5 shows the results of the ADAPT global climate change analysis on cereal production in 2025 with six different climate scenarios superimposed on the Business As Usual socio-economic scenario for technology, income and population to 2025. The results show that for the Sacramento (USA) and Rhine Basins, cereal production is enhanced by climate change under all climate change scenarios by as much as 40%. For the Volta, Zayandeh Rud, Walawe and Mekong Basins, all scenarios result in a decrease of cereal production compared to the current climate scenario. However, for the Walawe and Mekong Basins the magnitude of decreases are not large, while for the Volta and Zayandeh Rud Basins the impacts are significant, with up to 55% reduction in the Zayandeh Rud Basin. The Syr Darya Basin results show two scenarios enhancing production up to 5% and four scenarios show decreasing production up to 17%.

Conclusions

There are only two general conclusions that can be drawn from the results in Tables 12.4 and 12.5.

1. *Regional cereal prices decrease under climate change.* The results show that for all basins the price of cereals in 2025 goes down due to the favourable climatic conditions and CO_2 fertilization in some of the ADAPT basins. This leads to reduced world market prices, which affects regional prices.

2. *Regional cereal demand increases under climate change.* In most cases regional cereal demand increased, with the Rhine and Syr Darya being the exceptions.

The second conclusion above is quite interesting. Since many ADAPT regions experience reduction in production, cereal production goes down in all cases for the Volta, Zayandeh Rud, Walawe and Mekong Basins. The increase in demand is due to reduction in regional prices, caused by an increase in global production and a reduction in world market prices. Therefore, farmer incomes may be decreased by climate change, while consumers benefit from the lower prices, and the risk of hunger is reduced.

References

Alcamo, J., Henrichs, T. and Roesch, T. (2000) *World Water in 2025 – Global Modeling and Scenario Analysis for the World Water Commission on Water for the 21st Century*. Kassel World Water Series 2. University of Kassel, Kassel, Germany.

CSCDGC (2002) Plant growth data. Center for the Study of Carbon Dioxide and Global Change in Tempe, Arizona. Available at: http://www.co2science.org

Cosgrove, W.J. and Rijsberman, F.R. (2000) *World Water Vision: Making Water Everybody's Business*. Earthscan, London.

Droogers, P. and van Dam, J.C. (2004) Field scale adaptation strategies to climate change to sustain food security: a modeling approach across seven contrasting basins. IWMI Working Paper. International Water Management Institute, Sri Lanka.

FAO (2002) *World Agriculture: Towards 2015–2030*. Summary report. FAO, Rome.

Fischer, G., van Velthuizen, H.T., Shah, M.M. and Nachtergaele, F.O. (2003) *Global Agro-ecological Assessment for Agriculture in the 21st Century: Methodology and Results*. RR-02-002. Inter-national Institute for Applied Systems Analysis, Laxenburg.

Jones, M.H. and Curtis, P.S. (2000) Bibliography on CO_2 effects on vegetation and ecosystems: 1990–1999. Available at: http://cdiac.esd.ornl.gov/epubs/cdiac/cdiac129/cdiac129.html

New, M.G., Hulme, M. and Jones, P.D. (2000) Twentieth-century space–time climate variability. Part II: Development of 1901–1996 monthly grids of terrestrial surface climate. *Journal of Climate* 13, 2217–2238.

Rosegrant, M.W., Paisner, M.S., Meijer, S. and Witcover, J. (2001) *Global Food Projections to 2020: Emerging Trends and Alternative Futures*. International Food Policy Research Institute, Washington, DC.

Seckler, D., Barker, R. and Amarasinghe, U. (1999) Water scarcity in the twenty-first century. *Water Resources Development* 15, 29–42.

Yates, D. (1996) WatBal: an integrated water balance model for climate impact assessment of river basin runoff. *Water Resources Development* 12, 121–139.

13 Adaptation to Climate Change: a Research Agenda for the Future

JEROEN AERTS,[1] PETER DROOGERS[2] AND SASKIA WERNERS[3]

[1]Institute for Environmental Studies, Vrije Universiteit Amsterdam, Amsterdam, The Netherlands; [2]FutureWater, Arnhem, The Netherlands; [3]Climate Change and Biosphere Research Centre (CCB), Wageningen, The Netherlands

Adaptation to climate change increasingly receives attention in policy making as a complementary coping mechanism to mitigation (UNFCCC, 1997). Many adaptations have been developed in the past and throughout history people have adapted to changing or extreme climate conditions. However, water managers see themselves confronted with increasingly credible scientific information on the potential magnitude of climate change and climate variability and the vulnerability of water resources to its impacts. The urgency to take action is more apparent than ever, yet clear guidance on exactly how to respond to the challenge of climate change is lacking (Kabat and van Schaik, 2003).

The ADAPT project addressed this urgency and provided water managers across very different regions with information on possible impacts and a framework to develop and evaluate adaptations. More specifically, the main goal of the ADAPT project was to create a generic framework for river basins that allows for developing and assessing adaptation strategies to alleviate climate impacts on food and the environment.

Experiences Using the AMR Framework

The adaptation methodology for river basins (AMR) framework, roughly followed in the ADAPT project, puts stakeholders in a central role in the adaptation process in order to be sure that all existing adaptations are considered and that potential new adaptations are realistic and can be implemented. It has become clear from research, though, that more effort is required to improve our understanding of the human response to climate change, as it is likely that socio-economic systems respond most to extreme manifestations of climate change (Yohe and Dowlatabadi, 1999). Adaptation research needs to further concentrate on stakeholder analyses, including using socio-economic scenarios. It also became clear that huge knowledge gaps exist between policy makers/water managers and scientists who should inform water management about possible impacts. It has been shown that the AMR framework can help to bridge the gap between science and policy. This has been achieved by incorporating both scientific-based indicators and policy-oriented indicators (see basin Chapters 5–11). Also, the iterative character of AMR helps 'learning by doing' and supports the interaction between stakeholders from science and management.

Another important issue is the scale that adaptations target. Most adaptations in the ADAPT project focused on the regional (basin) scale. However, it is clear that most impacts (especially in developing countries) will have to be dealt with at the local (village) scale. Adaptation strategies in water management should increasingly be developed and implemented at the local level. This includes thorough connections to the potential of livelihoods and seeking combined poverty reduction and climate adaptation measures.

Furthermore, the costs of adaptation are scarcely studied and even less is known about the benefits of adaptation. Most studies focus on total damage costs – including adaptation – and not on avoided damages through adaptation (e.g. Zeidler, 1997). The ADAPT project has not focused on this subject specifically, although studies in the Rhine Basin, for example, point towards the beneficial effects of implementing adaptation measures. New water and adaptation research should address factors that affect adaptive capacity, such as: institutional capacity, wealth, planning time, scale, etc. (Tol *et al.*, 1998).

Finally, more research is required on the *timing* of adaptations. Burton *et al.* (1998) point out that adaptation in socio-economic sectors is easier when investments are connected to activities with a shorter product cycle. For example, different cropping methods can be adjusted every year. But a forest has a life-cycle of decades, and hence feedback mechanisms are more difficult to simulate in advance. Dams are even costlier to reconstruct in order to meet new climate conditions.

It is important to realize that adaptation measures are not exclusively related to climate change, but may also be beneficial given other internal (= manageable) and external (= less manageable) stressors. Examples of stressors that should be taken into account in the selection of adaptation strategies are: land use change, population growth, increased competition between sectors (urban, industry, agriculture, nature), power generation, trans-boundary water allocation, environmental concerns. The term 'No regret strategy' is often used to indicate that an adaptation to climate change will be beneficial to other policy goals, even if future climate change turns out to be less than expected.

Specific Conclusions and Suggestions

The adaptation methodology for river basins (AMR) framework

The AMR is the first adaptation framework for water managers and is based largely on existing decision and adaptation frameworks (OECD, 1993; Smit *et al.*, 1999; Barker, 2003). The challenge in the development of AMR framework is the difference in water resources characteristics, environmental controversies and socio-economic issues across river basins. In response, AMR offers an integrated approach, taking into account socio-economic and environmental aspects. A difficult issue remains in implementing the results of adaptation studies, since many transboundary river basins do not have a region-wide management institute with a mandate. Yet, from a water management perspective, a basin-wide approach is preferred for developing and evaluating adaptation strategies and the framework can still be used as a vehicle for discussions.

Integral impact assessment of climate change and climate variability across different scales

ADAPT aimed to assess the overall effect of climate change and climate variability on food security and the environment in selected river basins. To do so, ADAPT assessed the impact of global scenarios of changes in precipitation, temperature, CO_2 levels and socio-economic development on:

- agricultural production at the *field level*;
- water availability, food production and the environment at the *river basin level*; and
- food production, demand and trade at the *global level*.

It proved particularly difficult to consolidate the results from the different levels. No model or modelling framework exists that integrates knowledge from the field, basin and global levels towards food security and adaptation at the river basin scale. Communication between researchers working at the same level in very different basins proved often easier than between those working at different levels. This is partly due to the incompatible formalization of the issues at hand that each require their own experience and mind set.

More fundamental is that the modelling frameworks at the various levels are each geared towards their own representative indicators. Since the research and policy questions differ, the indicators do so as well, making study results hard to compare. An example that illustrates this point is that, at the field scale, yield is the common parameter that integrates the relevant impacts of climate change and climate variation. At the basin level, however, overall food production may be more appropriate, whereas from a global trade perspective, food demand is a central parameter.

Involving stakeholders at the local and regional levels

Adaptation to climate change and variability faces the inherent challenge that the climate is driven by complex global processes, but that ultimately adaptations have to be implemented at the local level to tackle the impacts of climate change on daily life. Participation of local actors is a prerequisite for a broadly supported local adaptation plan. At the same time, the sub-regional or basin level offers a better prospect for mobilizing policy interest, as this is the common scale for planning, corresponding to the policy makers' mandate. Science has thus far failed to effectively communicate its knowledge and insights on the global causes and local impacts of climate change and climate variability to policy makers and the public. Science and policy makers have only just started to involve stakeholders in their scientific approaches and adaptation policy development.

Adaptation to the impacts of climate change will be realized by: (i) increasing the resilience of communities through targeted discussions between stakeholders, (ii) strengthening research, knowledge and public awareness on climate change and water resources through learning and communication between researchers and other stakeholders, and (iii) further development of participatory approaches at regional and local levels in the field through integrated (impact) assessment and cost–benefit assessment.

Using GCM scenarios for regional modelling

The most widely used climate change projections are provided by the Intergovernmental Panel on Climate Change (IPCC, 2001). These projections are based on outputs of general circulation models (GCMs) and are used in many studies that assess impacts of climate change on hydrology. A major weakness of using GCM outputs for regional impact studies is that the spatial and temporal resolution of current climate model outputs is too coarse to be used directly in (regional) hydrological models.

To prepare the coarse GCM data for regional studies, downscaling techniques can be used. Four methods have been explored within the ADAPT project. The results for the baseline period (1961–1990) were compared in graphs and tables using statistical test parameters. It appears that each method generates substantially different results. The commonly employed downscaling technique that correct future projections based on annual and monthly mean averages perform relatively badly compared to methods that differentiate corrections across all cells. The best downscaling method (method 4, Chapter 2) corrects the GCM outputs using the variance of the measured data.

Environmental aspects

The ADAPT approach recognizes the environment as one of the key aspects in the search for coping strategies to climate change and increased climate variability. However, when using the AMR framework with stakeholders, it appears that environmental issues often play a secondary role in the process for formulating adaptation strategies. This holds especially for regions in developing countries, where socioeconomic problems are relatively urgent, compared to developed countries. A strong signal from policy to integrally address environmental, social and economic aspects in water management is sometimes lacking in developing countries. This is not surprising, since food security is already under threat and should be secured first. Nevertheless, ADAPT has shown that the sustainability aspect can be used to develop adaptation strategies that are both beneficial to the environment and to economic objectives.

Challenges for adaptation research in food security

The previous chapters made it very clear that climate change has positive and negative impacts on food security. The research agenda should therefore focus on alleviating the negative impacts and benefiting from the positive impacts. Regarding the latter, research should focus on CO_2 fertilization. Here, the two key research issues are:

- to study options for moving from C3 to C4 plants, which includes technical and socio-economic aspects; and
- to ensure that sufficient water, nutrients and managerial knowledge are available to profit from enhanced CO_2 levels.

To minimize the negative impact of climate change, the following research areas should be addressed in the near future:

- the options to enhance the buffer capacity of individual farmers: loaning system, crop differentiation;
- the options to enhance the buffer capacity of river basins/countries: water storage, food storage, trade;
- proper pre-defined planning mechanisms for water allocation under different scenarios of water availability; and
- planning mechanisms for the integration of adaptation strategies into management and (non-)climatic policy.

Obviously, these components should be studied in an integrated way, where the issues mentioned are relevant on a regional scale, but a different emphasis might be required depending on local conditions.

Conclusions

The following major suggestions and conclusions can be drawn.

1. Without adaptation, climate change and climate variability will have a profound impact on water resources across the basins studied in the ADAPT project.
2. Adaptation strategies now mainly target social and economic impacts of climate change, based on sound research of technical components. Therefore, it is recommended to focus more on sustainability in formulating adaptations, since this concept embraces both economic and environmental issues in the long term.
3. Downscaling GCM scenarios for use in regional modelling is still prone to uncertainty. It is recommended to use a variance correction over time (e.g. months) and not mean average annual corrections.
4. Adaptation research on food security should focus on taking advantage of the positive aspects of climate change (enhanced CO_2 levels) and minimizing the negative aspects (increased variability).
5. Solutions to alleviate the negative impacts of climate change will be found by: (i) increasing the resilience of communities through targeted discussions between stakeholders, and (ii) strengthening research, knowledge and public awareness on climate change and water resources through learning and communication between researchers and other stakeholders.

6. It proved particularly difficult to consolidate the results from the field, basin and global levels. AMR is an example of a methodology that helps to bridge gaps across scales and between scientists and policy makers. However, the scale aspect still deserves more attention.

7. Adaptation to the impacts of climate change will be realized by strengthening research, knowledge sharing and public participation in fields as policy planning and integrated (impact) assessment.

References

Barker, T. (2003) Representing global climate change, adaptation and mitigation. *Global Environmental Change* 13, 1–6.

Burton, I., Smith, J. and Lenhart, S. (1998) Adaptation to climate change: theory and assessment. In: Feenstra, J.F., Burton, I., Smith, J.B. and Tol, R.S.J. (eds) *UNEP Handbook on Methods for Climate Change Impact Assessment and Adaptation Strategies*. Institute for Environmental Studies, Vrije Universiteit, Amsterdam.

IPCC (2001) *Climate Change 2001 – Impacts, Adaptation and Vulnerability. Contribution of Working Group II to the Third Assessment Report of the Intergovernmental Panel on Climate Change*. McCarthy, J.J., Canziani, O.F., Leary, N.A., Dokken, D.J. and White, K.S. (eds) Cambridge University Press, Cambridge.

Kabat, P. and van Schaik, H. (2003) *Climate Changes the Water Rules: How Water Managers Can Cope with Today's Climate Variability and Tomorrow's Climate Change*. Dialogue on Water and Climate. Printfine, Liverpool.

OECD (1993) *OECD Core Set of Indicators for Environmental Performance Reviews*. Environment Monograph No. 83. OECD, Paris.

Smit, B., Burton, I., Klein, R. and Street, R. (1999) The science of adaptation: a framework for assessment. *Mitigation and Adaptation Strategies for Global Change* 4, 199–213.

Tol, R.S.J., Fankhauser, S. and Smith, J.B. (1998) The scope for adaptation to climate change: what can we learn from the impact literature? *Global Environmental Change* 8, 109–123.

UNFCCC (1997) *Kyoto Protocol to the United Nations Framework Convention on Climate Change*. FCCC/CP/L7/Add.1, 10 December 1997. United Nations, New York.

Yohe, G. and Dowlatabadi, H. (1999) Risk and uncertainties, analysis and evaluation: lessons for adaptation and integration. *Mitigation and Adaptation Strategies for Global Change* 4, 319–329.

Zeidler, R.B. (1997) Climate change vulnerability and response strategies for the coastal zone of Poland. *Climatic Change* 36, 151–173.

Index

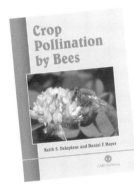
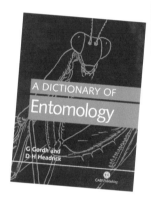
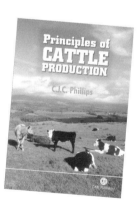